The Complete Book of

World War II USA POW & Internment Camp Chits

Prisoner of War Money in the United States

By

Dave Frank and David E. Seelye

COIN & CURRENCY INSTITUTE

82 Blair Park Road #399, Williston, Vermont 05495

(802) 878-0822 • Fax (802) 536-4787 • E-mail: mail@Coin-Currency.com

P·N·G
A. Friedberg

The Complete Book of World War II USA POW and Internment Camp Chits
Prisoner of War Money in the United States

Print edition:	ISBN 10: 087184-404-4	ISBN 13: 978-0-87184-404-0
E-book edition:	ISBN 10: 087184-006-5	ISBN 13: 978-0-87184-006-6

OTHER BOOKS PUBLISHED BY COIN & CURRENCY INSTITUTE

GOLD COINS OF THE WORLD

COINS OF THE BRITISH WORLD

APPRAISING AND SELLING YOUR COINS

SO-CALLED DOLLARS

AMERICA'S FOREIGN COINS

INVESTOR'S GUIDE TO UNITED STATES COINS

MODERN WORLD COINS

CURRENT COINS OF THE WORLD

COINAGE & CONFLICT

THE EARLY DATED COINS OF EUROPE: 1234–1500

THE COMPLETE CATALOG OF UNITED STATES LARGE SIZE STAR NOTES 1910–1929

A COLLECTOR'S GUIDE TO POSTAGE & FRACTIONAL CURRENCY

UNITED STATES PAPER MONEY ERRORS

Cover designed by Stacy Frank

A publication of
COIN & CURRENCY INSTITUTE

Printed in the United States of America

TABLE OF CONTENTS

FOREWORD

World War II is one of the most highly collected and studied areas in all of numismatics. There are literally hundreds of ways of delving into this period of history and each is important in its own right.

The study of how prisoners of war were treated has many facets, among them the ways they were able to function financially. There are tremendous variances to such a study, and each involved nation had its own methods of dealing with this problem. The United States chose a fairly unified system of providing ticket-like chits to P.O.W.'s in many camps located throughout the country.

When circumstances permit, the collecting instinct will take hold and the affected individual will begin the usual process of gathering examples and seeking information on whatever items he has found. Such was the case with Albert Pick, late author and originator of the *Standard Catalog of World Paper Money*. He was a P.O.W. in the United States, and he became interested in the chits of his and other camps- so much so that he came out with a listing, the first of its kind, in 1953. It was some years later, in 1970, that Al Donn compiled his data into a hard-cover book published by Krause Publications.

A third listing is contained in the second edition of Lance Campbell's P.O.W. tome published in 1993. Since then the collecting of these chits has become much more popular, to the extent that co-authors David Seelye and David Frank were spurred on to do further research into these intriguing issues. The result, I am told, will be nothing short of fabulous.

Apparently they have been successful in locating a great deal of new information and have combined that with all previously compiled data. All images including booklet covers will be in color for the first time; additionally, there will be a new numbering system and updated valuations.

It will become known as "The Chit Book"- so says co-author David Seelye. After hearing what this work will encompass, I can heartily agree with him. Both he and David Frank are already well-known specialists in the field, and their combined efforts have now produced a truly landmark reference.

Neil Shafer

PREFACE

The last use of Prisoner of War chits by POWs occurred sometime in 1946. That is when they returned to their home countries to resume their life after the war. Marriages took place, families were raised and memories of the war's experience went into scrapbooks and drawers. Many of the treasures that were hidden away are coming to light again. Unlike some other areas of numismatics, new discoveries of collectibles that were created over 70 years ago are occurring almost every month. New information is constantly being added to what is known. We must thank the pioneers of Prisoners of War chit collecting starting with Albert Pick, who himself was a German POW in the United States. From material he encountered in the camps and other information from fellow POWs, he authored *Die Kantineschecks der Deutschen Kriegsgefangenen in den U.S.A.* in 1953. Following Pick was another important researcher and author, Arlie Slabaugh, who published *Prisoner of War Monies and Medals* in 1965. Arlie was also a dealer who produced a monthly list of articles and chits for sale. Probably the best source on just US POW chits was written in 1970 by Albert I. Donn. The book was published by Krause Publications and titled *World War II Prisoner of War Scrip of the United States*. The Donn book remained the standard until Lance K. Campbell published the first edition of *Prisoner of War and Concentration Camp Money of the 20th Century* in 1989. A second edition with much more information was published in 1993. Other than a few pages in *World War II Remembered*, there has been nothing published since 1995.

Hopefully, this book will spark a new interest in collecting and researching.

This book would not have been possible without the contributions of Larry Smulczenski, Garry Arva, Gary Maltese, Greg Fedner, Neil Shafer, Joel Shafer, John Heim, Fred Schwan, Robbie Robinson, Joe Boling, and Joe Usibelli.

INTRODUCTION

The Second World War began on December 7, 1941 for the United States. Though the United States captured its first POW, a Japanese sailor at Pearl Harbor, the majority of the 425,000 POWs held in the US were captured beginning in 1942 in North Africa. The United Kingdom had been in the war since 1939 and was rapidly filling up camps set up in the UK, Africa, and India. The UK asked Canada and the United States to begin setting up camps and start accepting POWs in early 1942. Transportation was not a problem since troop ships carrying soldiers to the war zones were returning empty. The empty transport ships could be used to move prisoners of war from Europe, Africa, and the Pacific back to the United States. The United States agreed to house 150,000 POWs, however, by December 31, 1942 only 1,881 were held in US camps.

In September 1942, the Provost Marshal General submitted plans to house 50,000 prisoners of war, at first using unused camps in the southwest to house 75% of them. A list was prepared of permanent camps already in service, camps under construction, and authorized and temporary camps with the grand total expected to be 78,218. Next, camps were to be expanded in other commands to increase the total to be housed by another 100,000. The peak level of POWs in US camps was 425,871 in May 1945, with 700 camps in 46 states.

Prisoners were segregated by nationality and category. The categories were German Army anti-Nazi prisoners, the remaining German Army prisoners, German Navy anti-Nazi prisoners, the remaining German Navy prisoners, Italian prisoners, and Japanese prisoners. Russians captured in German uniforms were considered German prisoners of war.

International Law Concerning Treatment of POWs

The Geneva Prisoner of War and Red Cross conventions of 1929 laid the groundwork as to how prisoners of war should be treated in the future. The convention declared that all prisoners of war shall at all times be humanely treated and protected from acts of violence. They shall not be housed in war zones or areas of combat or bombardment. Differences of treatment are permissible only on the basis of military rank. Prisoners are only required to declare name, rank, and serial number. Enlisted men who are physically fit may be employed as workman in non-military related work. Farming, food processing, clothing plants, non-military truck and automobile repair, road and bridge construction, and flood control are examples. They could not work in the manufacture of munitions, military equipment, etc.

POWs that worked were paid according to the pay scale for their rank in their country or the country in which they were housed. In the United States the monthly pay for a private was $21 a month in 1941. This was roughly the equivalent of 80 cents per day, which is the amount the POWs were paid. In addition, each POW was given 10 cents a day gratuitously to purchase certain necessities. Enlisted men were required to work if they wanted to be paid; however officers were not required to work and still received their regular pay. If officers decided to work, they would receive added pay.

Some of the pay received by the POWs could be withdrawn from the fund for the purpose of purchasing Canteen Coupons (Chit booklets). The face value of the booklets could range from 20 cents to $20. Purchases of items could be made using the coupons (chits) at the camp canteen. Only the canteen clerk was allowed to remove the chit from the booklet to pay for a purchase. Examples of items that could be purchased are soap, shaving materials, stationary and postage, cigarettes and tobacco, 3.2% beer, toothpaste and brushes, candy, and soft drinks. All funds remaining in POW's account, plus the value of all remaining chits they turned in, were paid to the POW at the time of reparation in the form of a check. Many POWs did quite well on returning to their home country with a sizable sum compared to what the others in their country were left with.

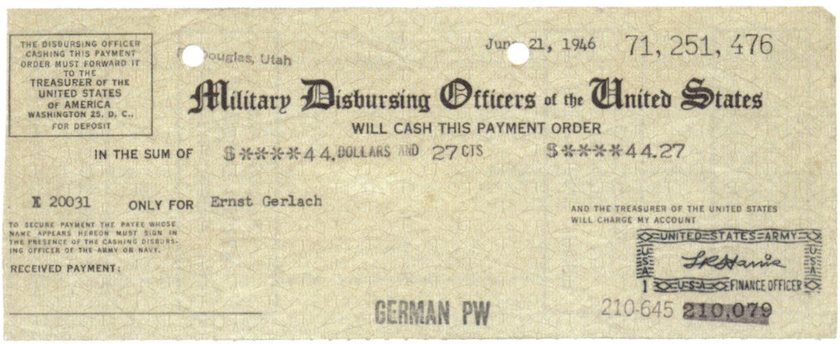
THE DISBURSING OFFICER CASHING THIS PAYMENT ORDER MUST FORWARD IT TO THE TREASURER OF THE UNITED STATES OF AMERICA WASHINGTON 25, D. C., FOR DEPOSIT

Douglas, Utah

Jun 21, 1946 71,251,476

Military Disbursing Officers of the United States

WILL CASH THIS PAYMENT ORDER

IN THE SUM OF $****44 DOLLARS AND 27 CTS $****44.27

X 20031 ONLY FOR Ernst Gerlach

TO SECURE PAYMENT THE PAYEE WHOSE NAME APPEARS HEREON MUST SIGN IN THE PRESENCE OF THE CASHING DISBURSING OFFICER OF THE ARMY OR NAVY.

RECEIVED PAYMENT:

AND THE TREASURER OF THE UNITED STATES WILL CHARGE MY ACCOUNT

UNITED STATES ARMY FINANCE OFFICER

GERMAN PW

210-645 210,079

The Camps

As stated earlier, approximately 700 POW camps existed in the US. Main camps were designated as base camps. The base camps administered not only their own operation but also the operation of outlying smaller camps, referred to as sub-camps or branch camps. Most likely, all base camps issued chits, even though some have yet to be discovered. It has been found that some branch camps also issued chits.

Base camps are identified by an asterisk (*) after the name of the camp in tables listing the camps in each state.

Printing of Chit Booklets

All issues of booklets were prepared by private contractors. The contracts were drawn up between the camp administration and the printers. Below is a table listing the known printers and some of the camps for which they issued scrip.

Camp	Booklet Denom.	Serial #		Printer	Printer Location	
AL-13	$5.00	21642C		Allison Coupon Co.	Indianapolis, IN	
AZ-10	UNK	A67294		Globe Ticket Co.	Los Angeles, CA	
AZ-20	$1.00	626A		The Kelsey Coupon Co.	Cincinnati, OH	
AZ-20	$1.00	110001H		The Kelsey Coupon Co.	Cincinnati, OH	
AR-8	$5.00	A 01990		Weldon, Williams & Lick	Ft. Smith, AR	
AR-16	$15.00	C 07866		Weldon, Williams & Lick	Ft. Smith, AR	
AR-16	$15.00	C 07957		Weldon, Williams & Lick	Ft. Smith, AR	
AR-16	$20.00	D 20003		Weldon, Williams & Lick	Ft. Smith, AR	
CA-51	$2.00	A04742		Globe Ticket Co.	Los Angeles, CA	July 31, 1945 - Sept. 30 1946
FL-4	$5.00	23159C		Southern Coupon Co.	Birmingham, AL	Feb 28 1946 - Apr 30
FL-5	$5.00	27A		Southern Coupon Co.	Birmingham, AL	
GA-8	$3.00	12510E		Southern Coupon Co.	Birmingham, AL	15 Jun 1945 - 25 Jun 1945
GA-14	$5.00	591K		Southern Coupon Co.	Birmingham, AL	31 May 1945 - 31 July
GA-14	$5.00	47738C		Southern Coupon Co.	Birmingham, AL	
HI-1	$2.00	5381B		Kansas Bank Note Co.	Fredonia, KS	Oct 16, 1945
IL-2	$4.00	14343C		Allison Coupon Co.	Indianapolis, IN	not valid after May 1, 1945
IL-2	$10.00	8006D		Allison Coupon Co.	Indianapolis, IN	
IL-2	$3.00	3 35209B		Allison Coupon Co.	Indianapolis, IN	31 Aug. 1945 - 31 Oct.
IL-2	$3.00	17666B		Allison Coupon Co.	Indianapolis, IN	
IL-2	$1.00	16211A		Allison Coupon Co.	Indianapolis, IN	
IL-2	80	48405		Allison Coupon Co.	Indianapolis, IN	
IL-2	80	30869		Allison Coupon Co.	Indianapolis, IN	
IL-3	40	2211G		Allison Coupon Co.	Indianapolis, IN	
IL-3	$3.00	6249C		Allison Coupon Co.	Indianapolis, IN	
IL-3	$1.00	8347A		Allison Coupon Co.	Indianapolis, IN	
IL-3	$5.00	34617E		Allison Coupon Co.	Indianapolis, IN	
IL-3	20	3152F		Allison Coupon Co.	Indianapolis, IN	
IN-2	$1.00	52934A	1537th	Allison Coupon Co.	Indianapolis, IN	
IN-2	$5.00	II 2908D	1537th	Allison Coupon Co.	Indianapolis, IN	
IN-2	$5.00	V	1560	Allison Coupon Co.	Indianapolis, IN	exp date Aug. 31, 1946
IN-2	$5.00	V	1560	Allison Coupon Co.	Indianapolis, IN	exp date Aug. 31, 1946
IA-1	$1.00	102501A		The Kelsey Coupon Co.	Cincinnati, OH	
IA-1	$1.00	90004A		The Kelsey Coupon Co.	Cincinnati, OH	
KS-8	$1.00	30876		The Kelsey Coupon Co.	Cincinnati, OH	
KY-3	$5.00	5529		Globe Ticket Co.	Los Angeles, CA	
LA-6	$5.00	C 04522		Weldon, Williams & Lick	Ft. Smith, AR	not valid before Oct. 31 - not valid after Dec. 31
MD-8	$10.00	19381B		Allison Coupon Co.	Indianapolis, IN	Oct 31 '45 - Dec 31 '45
MI-12	$1.00	11997A		Southern Coupon Co.	Birmingham, AL	
MI-12	$5.00	4 2589D		Southern Coupon Co.	Birmingham, AL	31 March 1946 - 31 May 1946

Camp	Booklet Denom.	Serial #		Printer	Printer Location	
MI-12	$5.00	8 11983N		Southern Coupon Co.	Birmingham, AL	31 July 1945 - 30 September 1945
MI-12	$5.00	8 161G		Southern Coupon Co.	Birmingham, AL	31 May 1945 - 31 July
MI-12	$5.00	8 919G		Southern Coupon Co.	Birmingham, AL	31 May 1946 - 31 July
MI-12	$5.00	8 1285G		Southern Coupon Co.	Birmingham, AL	31 May 1946 - 31 July
MI-12	$5.00	8 1387G		Southern Coupon Co.	Birmingham, AL	31 May 1946 - 31 July
MS-7	$1.00	25376A		The Kelsey Coupon Co.	Cincinnati, OH	
MO-33	$3.00	3192B		Southern Coupon Co.	Birmingham, AL	
MO-33	$3.00	5591B		Southern Coupon Co.	Birmingham, AL	
MO-33	$3.00	14035B		Southern Coupon Co.	Birmingham, AL	
NE-3	$10.00	C 8265		Army & Navy Literary	Washington,	15430
NE-3	$10.00	C 12428		Army & Navy Literary	Washington,	
NE-10	$5.00	44470BL		Southern Coupon Co.	Birmingham, AL	15461
NE-11	$1.00	21251A		The Kelsey Coupon Co.	Cincinnati, OH	
NE-11	$5.00	3501E		The Kelsey Coupon Co.	Cincinnati, OH	
NE-25	$3.00	19629B		Adams Bros. Salesbook	Topeka, KS	
NJ-1	$10.00	4501F	ISB	The Kelsey Coupon Co.	Cincinnati, OH	
NJ-3	$3.00	376C		The Kelsey Coupon Co.	Cincinnati, OH	
NJ-7	$3.00	112448A		Southern Coupon Co.	Birmingham, AL	
NM-16	$10.00	10001F		The Kelsey Coupon Co.	Cincinnati, OH	
NM-20	$3.00	1251C		The Kelsey Coupon Co.	Cincinnati, OH	
NY-12	$2.00	6751F		The Kelsey Coupon Co.	Cincinnati, OH	
NY-12	$2.00	6 24251F		The Kelsey Coupon Co.	Cincinnati, OH	31 October 1945 - 31 December 1945
NY-12	$2.00	8 28251F		The Kelsey Coupon Co.	Cincinnati, OH	31 December 1945 - 28 February 1946
NY-26	$1.00	4626A	ISB	The Kelsey Coupon Co.	Cincinnati, OH	
OK-1	$3.00	52126C		The Kelsey Coupon Co.	Cincinnati, OH	
OK-1	$5.00	23027E		The Kelsey Coupon Co.	Cincinnati, OH	
OK-1	$5.00	66188E		The Kelsey Coupon Co.	Cincinnati, OH	
OK-1	$5.00	64260E		The Kelsey Coupon Co.	Cincinnati, OH	
OK-9	$1.00	1251		The Kelsey Coupon Co.	Cincinnati, OH	not good after 25 November, 1945
OK-9	$3.00	126C		The Kelsey Coupon Co.	Cincinnati, OH	
OK-9	$3.00	1751C		The Kelsey Coupon Co.	Cincinnati, OH	
OK-13	$3.00	103292	Trade	Weldon, Williams & Lick	Ft. Smith, AR	
OR-2	$5.00	59044		Hancock Bros.	San Francisco,	
PA-3	$10.00	13698D		Allison Coupon Co.	Indianapolis, IN	June 30, 1945 - Aug. 30,
PA-25	$5.00	256E		The Kelsey Coupon Co.	Cincinnati, OH	
PA-25	$10.00	2501F		The Kelsey Coupon Co.	Cincinnati, OH	
RI-2	$3.00	630B		Southern Coupon Co.	Birmingham, AL	

Camp	Booklet Denom.	Serial #		Printer	Printer Location	
RI-4	$1.00	6 00646		Globe Ticket Co.	Boston, MA	
RI-4	$1.00	6 00649		Globe Ticket Co.	Boston, MA	
TN-1	$5.00	17751C		The Kelsey Coupon Co.	Cincinnati, OH	
TN-1	$5.00	41251D		The Kelsey Coupon Co.	Cincinnati, OH	
TN-1	$5.00	131251D		The Kelsey Coupon Co.	Cincinnati, OH	not valid after 25 June,
TN-1	$10.00	16001F		The Kelsey Coupon Co.	Cincinnati, OH	15279
TN-2	$1.00	3751B	intern-	The Kelsey Coupon Co.	Cincinnati, OH	
TN-2	$2.00	456B	POW	The Kelsey Coupon Co.	Cincinnati, OH	
TN-2	$10.00	1104F	POW	The Kelsey Coupon Co.	Cincinnati, OH	31 October 1945 - 31 December 1945
TN-5	$1.00	4001A	A.S.F.	The Kelsey Coupon Co.	Cincinnati, OH	
TN-5	$1.00	6501A	A.S.F.	The Kelsey Coupon Co.	Cincinnati, OH	not valid after 25 June,
TN-5	$1.00	10501A	A.S.F.	The Kelsey Coupon Co.	Cincinnati, OH	not valid after 25 June,
TX-15	$5.00	2501E		The Kelsey Coupon Co.	Cincinnati, OH	
TX-15	$5.00	5501E		The Kelsey Coupon Co.	Cincinnati, OH	
TX-16	$3.00	39278		Southern Coupon Co.	Birmingham, AL	Not good after 25 May
TX-16	$1.00	166B		The Kelsey Coupon Co.	Cincinnati, OH	expires 30 June 1946
TX-43	$5.00	5475E		Rand McNally & Com-	Chicago, IL	
TX-48	$0.50	3501N	Sample	Rand McNally & Com-	Chicago, IL	
TX-51	$1.00	37626A		The Kelsey Coupon Co.	Cincinnati, OH	
TX-60	$3.00	3251C	Intern-	The Kelsey Coupon Co.	Cincinnati, OH	
TX-60	$1.00	18626A	POW	The Kelsey Coupon Co.	Cincinnati, OH	
TX-60	$10.00	61001E	POW	The Kelsey Coupon Co.	Cincinnati, OH	
VA-4	$3.00	6626C		The Kelsey Coupon Co.	Cincinnati, OH	

Booklets

POW chits were issued in booklet form with strips of either four or five chits on a page. Five on a page is the normal amount. The cover of the booklet had a serial number, and each chit in that book would have the same serial number. Booklets were produced in a wide variety of denominations. These include 20 cents, 40 cents, 50 cents, 80 cents, $1, $2, $3, $4, $5, $6, $8, $10, and $20. The total of chits in the booklet would add up to the denomination of the booklet. Chits also come in a variety of denominations including 1 cent, 2 cents, 3 cents, 5 cents, 10 cents, 25 cents, 50 cents, $1, and $2. In addition to serial numbers, booklets and chits sometimes contain letters. These are referred to as block letters. They signify various print production issues. Therefore, a serial number without any block letter would be the first series for that camp printed. One with an A block letter would be next, followed by a third printing with a B and so on. Block letters can either precede or follow the serial number. Some chits have different printings symbolized by a large numeral, either preceeding the serial number or just inside the left edge of the chit. Holton, ME is an example of this type. Inside the booklet, after the cover but before the pages of chits, is generally a receipt page. This was usually signed by the POW confirming receipt of the booklet.

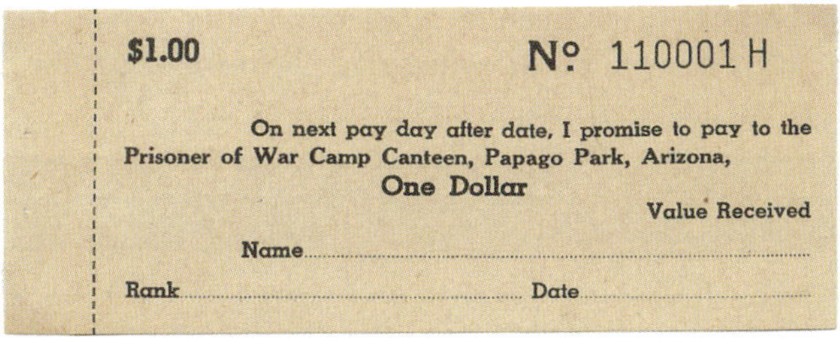

The cover of the booklet had the name of the camp, the value of the booklet, sometimes how long the chits were valid, the name of the POW, his rank and organization, a signature of the canteen officer, and the date.

Pattern of Chits in Booklet Denominations

Face Value	Chits Included
$1	1, 5, 10 cents
$2	1, 5, 10, 25 cents
$3	1, 5, 10, 25 cents
$5	1, 5, 10, 25, 50 cents
$10	1, 5, 10, 25, 50 cents, $1
$20	1, 5, 10, 25, 50 cents, $1

The above table is a generalization of the denominations of chits that will be found in booklets of that face value. There are other denominations that are occasionally found in booklets, such as 2 cents, 3 cents, and 2 dollars. Also, as stated earlier, some booklets have 4 to a page while most have 5. In addition, some of the pages of 5 may have a blank or spacer to make the total of all of the chits equal the face value of the book. The contents of several booklets are given below to show the number of chits per denomination that appear in different face value booklets.

Camp	Booklet Value	Chit sheets
Kilmer	$2	1 x 4 at 25¢
		1 x 4 at 10¢
		2 x 4 at 5¢
		5 x 4 at 1¢
Kilmer	$3	1 x 4 at 25¢
		3 x 4 at 10¢
		3 x 4 at 5¢
		5 x 4 at 1 ¢
Pryor	$5	2 x 5 at 25¢
		3 x 5 at 10¢
		3 x 5 at 5¢
		1 x 5 at 2¢
		3 x 5 at 1¢
Crossville	$5	2 x 5 at 25¢
		3 x 5 at 10¢
		3 x 5 at 5¢
		1 x 5 at 3¢
		2 x 5 at 1¢
Camp White	$5	2 x 5 at 25¢
		1 x 4 at 10¢ plus blank
		3 x 5 at 10¢
		2 x 5 at 5¢
		2 x 5 at 1¢
Port Terminals	$5	2 x 5 at 25¢
		3 x 5 at 10¢
		3 x 5 at 5¢
		5 x 5 at 1¢
Douglas	$1	1 x 5 at 10¢
		1 x 5 at 5¢
		5 x 5 at 1¢

This sample of the contents of a few booklets will show that the higher denomination of the chit, the fewer are present in the booklets.. That is one reason why 1 cent chits are the most common followed by 5 and 10 cent chits. The 25 and especially 50 cent chits are the scarcest. Chits in 2 cent, 3 cent and $1 denominations are very difficult to locate. Only one $2 chit is known. Another reason that the lower denominations are easier to find is that POWs returning home were more likely to save a lower denomination as a souvenir and cash in the higher ones.

Other Chits Used

In addition to booklets and chits that specifically are printed or stamped POW, P.O.W., PW, or Prisoner of War, sometimes regular base canteen, NCO, or Merchandise Coupons were used. This is known because many of these have been found in POW's scrap books or have POW written on the back of the chit. Chits were also issued for Internment Camps and Italian Service Units. These will be discussed later in those specific chapters.

Printing Styles

Chits were produced in a variety of printing styles. There are some standard types that will be seen for many camps. These were generally produced by the same printer. Some camps may have three or more different types. Sometimes the denomination could be on the right or left. The wording may be different such as POW Exchange or POW Canteen. This makes it both fun and challenging to collect not only the camp but the different types for an individual camp. Many collectors also collect by different serial numbers for the individual camp chits. That is why there is a table listing all of the serial numbers that the authors have discovered. The list will be expanded as collectors submit new serial numbers. Space is provided in this section for collectors to log their personal discoveries.

Book Numbering System

In order to catalog the chits, this book has created a unique numbering system. The goal is to provide some type of identification about where and in what camp the chit was used. The catalog number begins with the two letter abbreviation of the state where the camp is located. The next number refers to the base camp listed in the table for that state. The third number is the chit type from that camp. Some camps have only one type but others could have two, three, four, or five. Lastly, the final number refers to the denomination of the chit.

Table of Chits with Values

Each camp will have a table that lists the catalog number, the denomination, paper color, ink color, serial number color, known block letters, and value in circulated and uncirculated condition. Determining the paper color can be difficult due to age and fading of the paper, in addition to subjectivity between collectors as to what color is represented.

The designation "RNK" in the tables that follow indicate that a chit is reported, but not known.

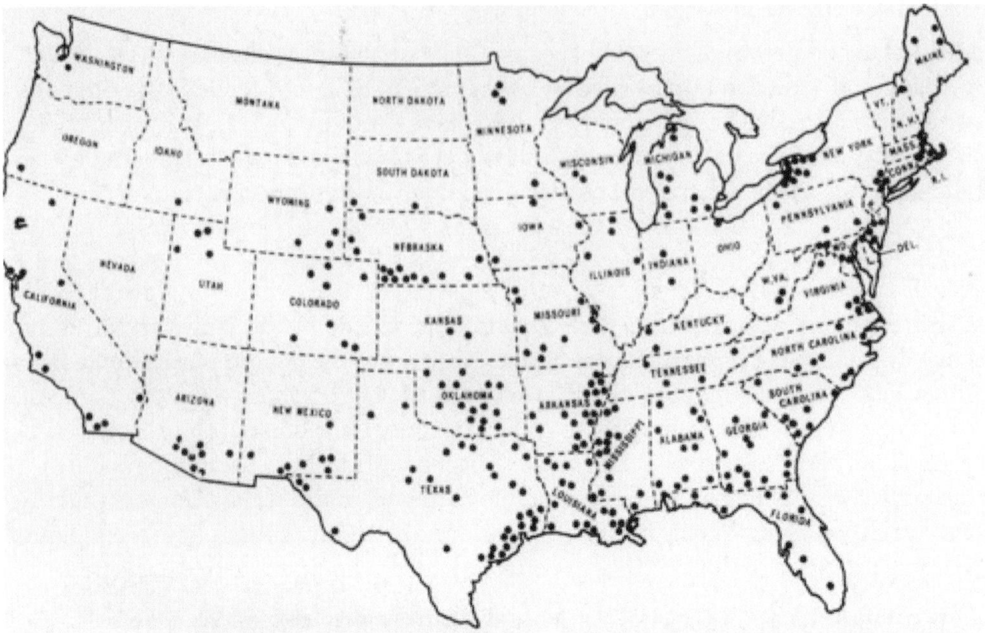

Distribution of the major POW camps across the United States as of June 1944 (U.S. Army Photo)

POWs arriving at the Port of Boston

POWs on the way to a work detail at Atterbury, Indiana

ALABAMA

Alabama had four POW base camps located within the state—Aliceville, Ft McClellan, Opelika, and Camp Rucker.

	Camp	Base Camp
1	Abbeville	Camp Rucker
2	*Aliceville	
3	Andalusia	Shelby, MS
4	*Camp Rucker	
5	*Camp Sibert	
6	Chatom	Shelby, MS
7	Clanton	Fort McClellan
8	Clio	Camp Rucker
9	Daleville	McCain
10	Dothan	Clinton
11	Evergreen	Camp Rucker
12	Foley	Camp Rucker
13	*Fort McClellan	
14	Geneva	Shelby, MS
15	Greenville	Camp Rucker
16	Huntsville Ars.	Camp Forrest, TN
17	Jackson	Shelby, MS
18	Loxley	Shelby, MS
19	Luvern	Fort McClellan
20	Montgomery H&R	Camp Rucker
21	Northington GH	Aliceville
22	Oneonta	Camp Forrest, TN
23	*Opelika	
24	Troy	Fort McClellan

Aliceville POW Base Camp

Aliceville was located in the West Central part of the state near the Mississippi river. The camp was built in 1942 and was one of the largest POW camps in the southeastern part of the United States. Most of the prisoners were from Rommel's Africa Korps. After the war, the camp's buildings and land were sold at auction to the city and private individuals.

AL-2-1-1

AL-2-1-2

AL-2-1-5

AL-2-1-10

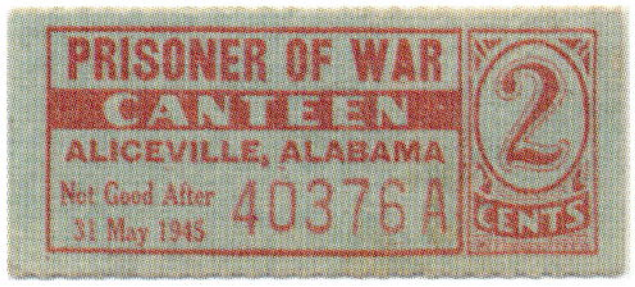

AL- 2-3-2

AL-2-3-10

No.	Denomination	Paper	Ink	Serial	Block	Circ.	Unc.
	Prisoner of War Canteen						
AL-2-1-1	1 cent	buff	blue	red	5 G	200	400
AL-2-1-2	2 cents	blue	blue	red	5 G	150	250
AL-2-1-5	5 cents	orange	blue	red	5 G	150	250
AL-2-1-10	10 cents	pink	blue	red	5 G	175	300
	Prisoner of War Camp Canteen						
AL-2-2-1	1 cent	yellow	blue	red	G		RNK
AL-2-2-2	2 cents	blue	blue	red	G		RNK
AL-2-2-5	5 cents	orange	blue	red	G		RNK
AL-2-2-10	10 cents	pink	blue	red	G		RNK
	Dated May 31, 1945						
AL-2-3-2	2 cents	blue	red	red	A	175	350
AL-2-3-10	10 cents	pink	red	red	A	200	400
	Dated July 31, 1945						
AL-2-4-1	1 cent	yellow	blue	red			RNK
AL-2-4-2	2 cents	blue	blue	red			RNK
AL-2-4-5	5 cents	orange	blue	red			RNK
AL-2-4-10	10 cents	pink	blue	red			RNK

Ft. McClellan POW Base Camp

Ft. McClellan has a proud and fascinating history that dates back to the First World War. The War Department established Camp McClellan on July 18, 1917. The camp was named in honor of Major General George B. McClellan, General-in-chief of the U.S. Army from 1861-1862. McClellan was also Governor of New Jersey from 1878-1881. Although it is unusual for a Southern fort to be named for a Northern general, there are strong indications that McClellan's name was a logical choice. Camp McClellan was a mobilization camp used to quickly train men for WWII. General McClellan is credited with the training and mobilization of the Army of the Potomac during the Civil War. Most of the soldiers arriving at the camp were from the North. In fact, the first group to train at Camp McClellan were from McClellan's home state of New Jersey.

A 3,000 capacity POW area was built during 1943 when Fort McClellan became the temporary home for 2,758 captured enemy soldiers. This facility continued to hold captives until June of 1946. A memorial cemetery located near the western corner of the post is the final resting place for 26 German and three Italian POWs who died in captivity.

In 1995, the Base Realignment and Closure (BRAC) Commission voted to permanently close Fort McClellan. The official closing ceremony ending Fort McClellan's illustrious past was held on 20 May, 1999.

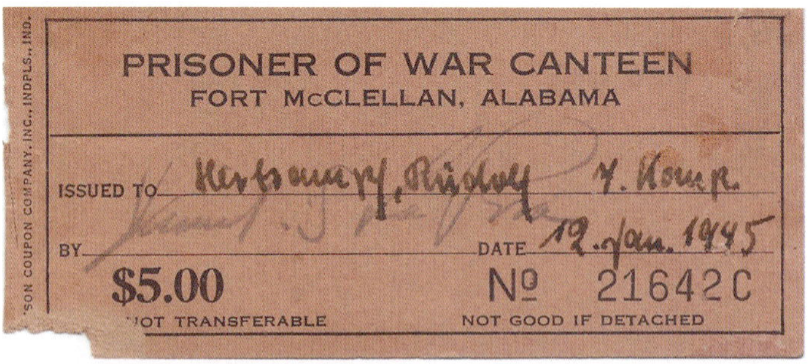

AL-13-1-5

AL-13-2-1

No.	Denomination	Paper	Ink	Serial	Block	Circ.	Unc.
AL-13-1-1	1 cent	white	blue	red	None	150	175
AL-13-1-5	5 cents	pink	blue	red	C	150	175
AL-13-1-10	10 cents	white	blue	red		150	175
	Post Exchange						
AL-13-2-1	1 cent	gray	blue	red	B	200	250

Opelika POW Base Camp

Opelika was a large main POW camp holding up to 3,000 prisoners and was built on the outskirts of Opelika, AL. Opelika is the county seat of Lee County. Opelika, meaning "big swamp", draws its name from the Creek Indian language. Very little is known about the camp other than it opened in June 1943, had a peak population of 2,772 prisoners, and closed in December 1945. Two series of chits are known from this camp, one entitled "Internment Camp", and the other "Prisoner of War Camp."

AL-23-1-1

AL-23-1-10

No.	Denomination	Paper	Ink	Serial	Block	Circ.	Unc.
AL-23-1-1	1 cent	rose	black	red	B	175	350
AL-23-1-5	5 cents	pink	black	red	C	175	350
AL-23-1-10	10 cents	white	black	red		175	250
	Post Exchange						
AL-23-2-1	1 cent	gray	blue	red	B	100	150

Camp Rucker POW Base Camp

During the 1930's, a 35,000 acre tract of land in southeastern Alabama was purchased by the Federal Government and converted into a wildlife refuge named the Pea River Land Use Project. In July 1941, the War Department selected the lands of the Pea River Project to be used as an infantry training camp. It was necessary to acquire an additional 30,000 acres of land, mostly owned by the federal government and the Atlantic Coast Line railroad between Newton and Enterprise.

In January 1942, a few weeks after the Japanese attack on Pearl Harbor, the U.S. Army Corps of Engineers completed plans for the 4,600 acre cantonment area of the camp. These plans called for 1,500 buildings, streets, utilities, wells, railroad side tracks, and other facilities. An army of 12,000 civilian workers tackled the job of constructing Camp Rucker. In less than 120 days the job was completed.

Camp Rucker takes its name from Colonel Edmund W. Rucker, an outstanding officer of the Confederate forces during the Civil War.

In June of 1943, the first German POWs were interned at Camp Rucker in stockades near the railroad, east of the warehouse area, on the southern edge of the post. There were 1,718 prisoners at Camp Rucker, mostly German but a few Italians prior to its closing in March 1946.

Camp Rucker was inactive between March 1946 and August 1950, when it was reactivated for the Korean hostilities. The Army Aviation School moved to Rucker from Fort Sill, Oklahoma in 1954. Today, Fort Rucker is the center for the U.S. Army Aviation School and the "home" of Army Aviation.

AL- 4-2-1

AL-4-4-2a

AL-4-4-2b

AL-4-3-5

No.	Denomination	Paper	Ink	Serial	Block	Circ.	Unc.
	Series 1						
AL-4-1-1	1 cent	white	blue	red			RNK
AL-4-1-2a	2 cents	red	blue	red			RNK
AL-4-1-2b	2 cents	white	blue	red			RNK
AL-4-1-5	5 cents	maroon	blue	red			RNK
	Series 2						
AL-4-2-1	1 cent	tan	blue	red	B	175	350
AL-4-2-5	5 cents	maroon	blue	red			RNK
AL-4-2-10	10 cents	bl/green	blue	red			RNK
	Barber ticket Fantasy						
AL-4-3-5	5 cents	tan	red	red		?	

	Fantasies						
AL-4-4-2a	2 cents	tan	black	none		?	
AL-4-4-2b	2 cents	tan	red	red	A	?	

Camp Sibert

A sub-camp for which, until recently, no chits were known.

AL-5-1-1

No.	Denomination	Paper	Ink	Serial	Block	Circ.	Unc.
AL-5-1-1	1 cent	rose	black	red	E	175	350

ALASKA

Excursion Inlet, located about 38 miles west of Juneau on the southern border of Glacier National Park, was the only POW camp in Alaska. During the Alaska Campaign, and because of the Japanese attack on Dutch Harbor, the government built an airfield and associated buildings at Gustavus, just to the west of Excursion Inlet. A highly secret project to build a huge supply depot was started there in August 1942 and completed in November 1943.

This terminal consisted of 800 buildings, three large docks, and covered 600 acres. However, by the time it was completed, the war effort moved further into the South Pacific, where the facilities were used only for a few months. Sometime after June of 1945, 700 German prisoners were shipped to Excursion Inlet to dismantle the terminal. Some of the material that was salvaged was used to rebuild the village of Hoonah, which had suffered a devastating fire the previous year. The camp opened in August of 1945 as a sub-camp of Fort Lawton, Washington, was upgraded to a base camp in October 1945, and closed in January 1946. Ocean Beauty Seafood currently operates one of the largest canneries in the world at this location.

The first chit seen from this location was discovered in 2013. There is also a picture showing a "KANTINE" operated by the Germans with a chit book in the hands of a POW.

	Alaska
1	*Excursion Inlet

AK-1-1-1

No.	Denomination	Paper	Ink	Serial	Bloc	Circ.	Unc
AK-1-1-1	1 cent	tan	black	red	A	500	750

ARIZONA

There were six base camps in Arizona—Davis Monthan, Florence, Papago Park, Pima, Colorado River Relocation Camp, and Gila River Relocation Camp.

	Camp	Base Camp
1	Casa Grande	Florence
2	Continental	Papago Park
3	Cortaro	Florence
4	Cotton Center	Papago Park
5	*Davis-Monthan	
6	Duncan	Papago Park
7	Eleven Mile Corner	Florence
8	Elroy1	Papago Park
9	Elroy11	Florence
10	*Florence	
11	Litchfield	Papago Park
12	Luke Field	Florence
13	Maricopa	Flrence
14	Mesa	Papago Park
15	McDowell Road	Florence
16	Mt. Graham	Florence
17	Navajo Ord Dep	
18	Parker	Florence

19	*Papago Park	
20	Pima	
21	Queen Creek	Papago Park
22	Roll	Florence
23	Safford	Florence
24	Yuma	Papago Park

Davis Monthan POW Base Camp

Davis Monthan is located on the southwest edge of Tucson. It was named in honor of two Tucson aviators who lost their lives in aircraft accidents after World War I. Lt. Samuel H. Davis died in a Florida aircraft crash in 1921, while Lt. Oscar Monthan lost his life in the Hawaii crash of a Martin bomber in 1924. The POW camp was physically on the Davis Monthan Air Base. The camp operated from May 1945 to March 1946, with a population of 423 German POWs. Only one issue of chits is known.

AZ-5-1-1

AZ-5-1-5

No.	Denomination	Paper	Ink	Serial	Block	Circ.	Unc.
AZ-5-1-1	1 cent	yellow	blue	red	A	25	40
AZ-5-1-5	5 cents	orange	blue	red	A	35	55
AZ-5-1-10	10 cents	pink	blue	red	A	55	85
AZ-5-1-25	25 cents						RNK

Florence POW Base Camp

The town of Florence is 45 miles southeast of Phoenix with the camp just north of town. The town and camp are believed to be named after a sister or daughter of one of the early city founders. The POW camp, built in 1942-43 was the largest all-new compound in the United States. Originally, the camp housed only Italian enlisted men, but, with the defeat of Italy, the prisoners who renounced the Fascist government were sent to service units at other locations. The invasion of Normandy saw the influx of German POWs to Florence which endured a population of 9,346 prisoners. Both Italians and Germans worked in agriculture during their incarceration.

AZ-10-1-1

AZ-10-1-5

AZ-10-2-1a

AZ-10-2-1b

AZ-10-2-5

AZ-10-2-10

AZ-10-3-1

No.	Denomination	Paper	Ink	Serial	Block	Circ.	Unc.
	P.O.W. Canteens						
AZ-10-1-1	1 cent	buff	black	black		150	250
AZ-10-1-5	5 cents	pale green	black	black		175	300
AZ-10-1-10	10 cents	orange	black	black		200	350
AZ-10-1-25	25 cents	yellow	black	black			RNK
AZ-10-1-50	50 cents	yellow	black	black			RNK
	PW Canteens						
AZ-10-2-1a	1 cent	pink	black	black	C	150	250

AZ-10-2-1b	1 cent	pink	black	black	B	150	250
AZ-10-2-5	5 cents	pale blue	black	black	B	175	300
AZ-10-2-10	10 cents	red	black	black	B	200	350
	Camp Exchange						
AZ-10-3-1	1 cent for U.S. Military Personal	gray	black	black	A	150	300

Papago Park POW Base Camp

Papago Park was located eight miles east of downtown Phoenix at the site of a pre-war CCC camp. The name "Papago" comes from a derogatory slang expression for the Toghono O'odhom Indians which likely means tepary-bean eaters. The camp housed mainly German naval prisoners, including a number of U-boat crews. It is well known for the three foot diameter tunnel dug by a group of prisoners for 178 feet, 8 to 14 feet below the surface. The escape plan was masterminded by a 43 year old U-boat captain by the name of Jurgen Wattenberg. His plan was to build the tunnel out to the Cross Cut Canal, then float down the Salt River, to the Gila River, and finally down the Colorado River into Mexico. Wattenberg had somehow obtained the names and addresses of people in Mexico who would help them get back to Germany. The tunnel was started in September behind the bathhouse. It ran under two camp fences, a drainage ditch and a road. Tools for this dig were obtained under the ruse that the POWs wanted to build a volleyball court for their recreation. Dirt from the tunnel was scattered around the camp and in the area of the volleyball court construction.

The tunnel was completed in December of 1944. Wattenberg ordered the men of an adjacent area of the camp to throw a boisterous party on the night of December 23rd to cover the noise of the escape. Twelve officers and 13 enlisted submariners went through the tunnel in groups of two or three during the wee hours of Christmas Eve 1944. The escapees became frustrated when they found that the rivers that they were about to use to float into Mexico and freedom were nothing more than dried up river beds and, at best, mud puddles. In no time at all, hungry, wet and cold escapees began to turn themselves into local American authorities. Wattenberg actually evaded capture for a month by traveling northwest toward the mountains. On Saturday, January 27, 1945, he decided to go to Phoenix and get a job as a dishwasher in order to earn enough money to buy a train ticket or possibly just hop aboard a freight train. He was tired, became careless and fell asleep for a couple of hours in a small hotel lobby. He woke after midnight and noticed the bellhop watching him. He hurriedly left but asked a man in the street for directions to the train station. The man hailed a policeman, saying that he was asked directions by an individual with a heavy German accent. Wattenberg's evasion had ended and he was returned to Papago Park. He was fed a Sunday meal of chicken and vegetables, but was then placed on bread and water for two weeks.

Papago Park opened as a POW camp in September 1943 and closed in April 1946. Peak population varied between 1,700 and 2,500 prisoners.

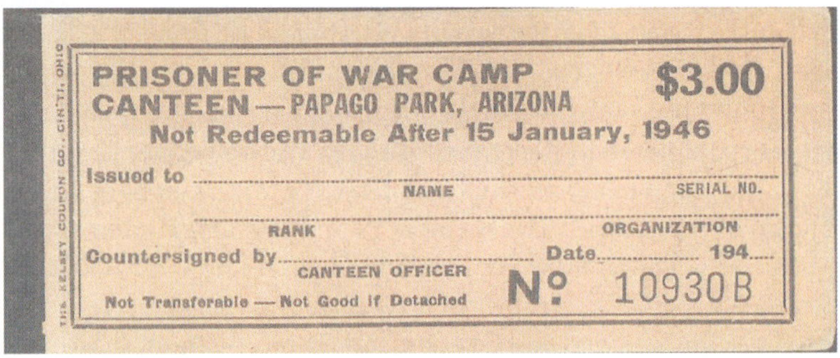

AZ-19-1-1a

AZ-19-1-1d

AZ-19-1-2a

AZ-19-1-2b

AZ-19-1-5a

AZ-19-1-5b

AZ-19-1-5c

AZ-19-1-10b

AZ-19-1-10c

AZ-19-1-10d

AZ-19-1-10f

AZ-19-1-25

No.	Denomination	Paper	Ink	Serial	Block	Circ.	Unc.
AZ-19-1-1a	1 cent	gray	blue	red	A	25	40
AZ-19-1-1b	1 cent	green	blue	red	A	25	40
AZ-19-1-1c	1 cent	manila	blue	red	H	25	40
AZ-19-1-1d	1 cent	pink	blue	red	J	25	40
AZ-19-1-2a	2 cents	salmon	blue	red	G	100	200
AZ-19-1-2b	2 cents	white	blue	red	D	100	200
AZ-19-1-5a	5 cents	yellow	blue	red	A	35	55
AZ-19-1-5b	5 cents	green	blue	red	G	35	55
AZ-19-1-5c	5 cents	orange	blue	red	J	35	55
AZ-19-1-5d	5 cents	salmon	blue	red	G	35	55
AZ-19-1-10a	10 cents	pale green	blue	red	A	55	85
AZ-19-1-10b	10 cents	orange	blue	red	A	55	85
AZ-19-1-10c	10 cents	green	blue	red	D	55	85
AZ-19-1-10d	10 cents	rose	blue	red	G	55	85
AZ-19-1-10e	10 cents	buff	blue	red	H	55	85
AZ-19-1-10f	10 cents	white	blue	red	J	55	85
AZ-19-1-25	25 cents	red	blue	red	E	90	150

Pima POW Base Camp

Very little is known about the Pima camp. It appears to have been located physically within the Papago Park compound from January 1945 to February 1946 with only 102 occupants. This was possibly a restricted site harboring a special category prisoner. High ranking officers, political prisoners, trouble makers, or maybe even

Japanese prisoners could have been interned at this site. This information is not known at this time. Since a peak population of only 102 individuals is recorded for this camp, it lends credence to the supposition that this was a selective group of prisoners. No chits are known from this camp, but because of the proximity of Papago Park, it might be assumed that Papago chits were used instead of a separate issue for this small population.

Colorado River Relocation Camp

This was one of the first camps built to temporarily house ethnic Japanese evicted from the west coast under the volunteer relocation program until they found work further to the east. It opened in March 1942, approximately three miles south of Poston, near the Southern California eastern border. The voluntary program was a total failure, so the camp was turned into a permanent relocation center. It was divided into three compounds that would house 20,000 people. The three compounds were known to the internees as "Roasten", "Toastin", and "Dustin" because of the hot and dry weather conditions. The camp never reached its capacity but did house a peak population of 17,800. Since it was located on an Indian Reservation, it was managed by the Bureau of Indian Affairs instead of the War Relocation Authority. All other relocation camps were administered by the WRA. No chits or tokens are known from this location. It closed in November of 1945.

Gila River Relocation Camp

This camp was located about 50 miles southeast of Phoenix near the small town of Rivers. This center was divided into two camps about three miles apart—Butte camp to the west and Canal camp to the east, each to accommodate 5,000 internees. This camp opened in May 1942 and was considered one of the least oppressive camps in the WRA system. However, conditions were harsh in the semi-desert environment, with scorpions and rattlesnakes a common problem. The camp exceeded its capabilities in late 1944 when over 2,000 internees from the Jerome, Arkansas camp were shipped to Gila River, swelling its population to 13,348. No chits are known from this camp. It closed in November 1945.

ARKANSAS

	Camp	Base Camp
1	Altheimer	Monticello
2	Bassett	Camp Joseph T. Robinson
3	Blytheville AAF	Camp Joseph T. Robinson
4	*Camp Chaffee	
5	*Camp Joseph T. Robinson	
6	Crawfordsville	Camp Joseph T. Robinson
7	Criderfield	Monticello
8	*Dermott	
9	Earle	Camp Joseph T. Robinson
10	Elaine	Camp Joseph T. Robinson
11	Grady	Monticello
12	Grider Field	Dermott
13	Harrisburg	Camp Joseph T. Robinson
14	Hot Springs	Camp Joseph T. Robinson
15	Hughes	Camp Joseph T. Robinson
16	*Jerome	
17	Jonesboro	Camp Joseph T. Robinson
18	Kaiser	Camp Joseph T. Robinson
19	Knobel	Camp Joseph T. Robinson
20	Lake Village	Dermott
21	Luxora	Camp Joseph T. Robinson
22	Magnolia	Monticello
23	Marked Tree	Camp Joseph T. Robinson
24	Montecello	
25	Murfreesboro	Camp Joseph T. Robinson
26	Newport	Camp Joseph T. Robinson
27	Osceola	Camp Joseph T. Robinson
28	Pine Bluff	
29	Russellville	Camp Chaffee
30	St. Charles	Camp Joseph T. Robinson

31	Simsboro	Camp Joseph T. Robinson
32	Stuttgart AAF	Camp Joseph T. Robinson
33	Turell	Camp Joseph T. Robinson
34	Victoria	Camp Joseph T. Robinson
35	Walnut Ridge	Camp Joseph T. Robinson
36	West Helena	Camp Joseph T. Robinson
37	Wynne	Camp Joseph T. Robinson

Chaffee POW Base Camp

This camp, located on the southeast edge of Fort Smith, Arkansas, was named for Major General Adna R. Chaffee Jr. who, as an artillery officer in Germany during WWI, realized that the Calvary was outmoded and advocated the use of tanks. The POW camp had a peak population of 3,002 German prisoners. It opened in June of 1943 and closed in July 1946.

AR-4-1-2

No.	Denomination	Paper	Ink	Serial	Block	Circ.	Unc.
AR-4-1-2	2 cents	pink	black	red		200	400

Dermott POW Base Camp

After the close of the Jerome Relocation Camp for ethnic Japanese internees on June 30, 1944 (see Jerome), this location was used as a POW Camp for German prisoners beginning in November 1944. On January 1, 1945, the camp was renamed for the small town of Dermott to the north of Jerome. No reason is known for the name change. The population of the camp swelled to 5,402. One chit is known for the camp with the name of Dermott. Possibly the chits for Jerome continued to be used even after the name change until supplies were exhausted. The camp closed in March 1946.

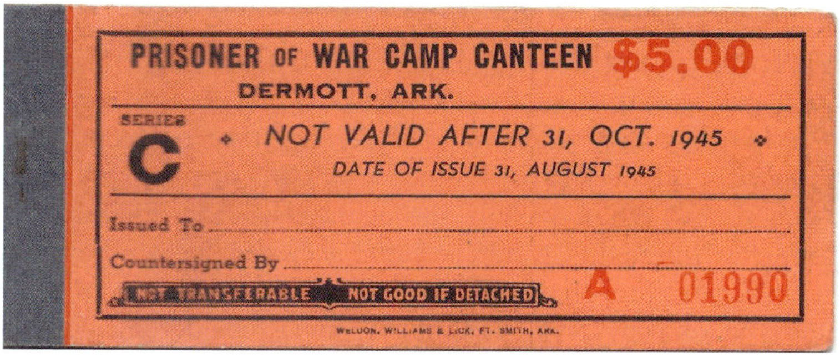

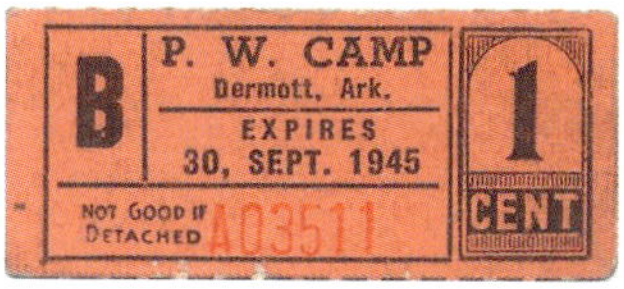

AR-8-1-1

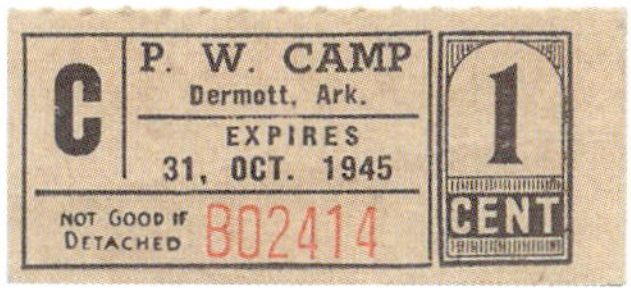

AR-8-2-1

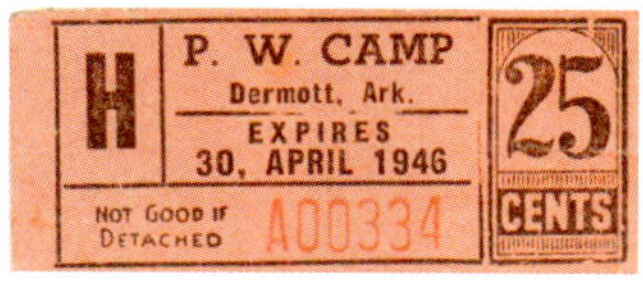

AR-8-3-25

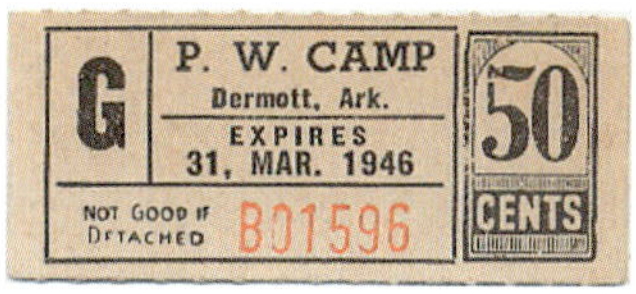

AR-8-4-50

No.	Denomination	Paper	Ink	Serial	Block	Circ.	Unc.
	Sept. 30, 1945						
AR-8-1-1	1 cent	orange	black	red	A	150	300
	31-Oct-45						
AR-8-2-1	1 cent	tan	black	red	B	150	300
	Mar. 31, 1946						
AR-8-3-25	25 cents	orange	black	red	A	175	350
	Apr. 30, 1946						
AR-8-4-50	50 cents	tan	black	red	B	200	400

Jerome POW Base Camp

Two nearby communities in southeast Arkansas, Jerome and Rohwerm, were allocated to house relocation camps for ethnic Japanese. These two camps were the easternmost sites chosen for this purpose. State officials had offered the land to the War Relocation Authority (WRA), hoping that land would be turned into valuable farmland by the internees since loggers had stripped the land of timber. The camp was built by a Dallas firm similar to other relocation camps to house approximately 8,000 people. The first ethnic Japanese began arriving in October 1942 and reached a peak population of 8,497. The Japanese were not welcome in Arkansas, a state well known for its history of segregation. When authorized to be in the local communities, the ethnic Japanese were required to use the rest rooms and drinking fountains reserved for African-Americans. When possible, the Japanese would work hard to find jobs outside of Arkansas where they would be better accepted. Jerome was the first Relocation Center to close in June 1944, with the remaining internees transferred to other locations. With the influx of German captives from Europe, Jerome was redesignated as a POW camp in November 1944 and renamed Dermott. (See Dermott)

AR-16-1-1

AR-16-1-5

AR-16-1-10

AR-16-1-25

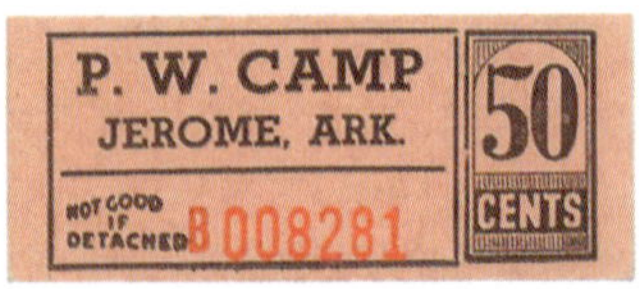

AR-16-1-50

AR-16-1-$1.00

No.	Denomination	Paper	Ink	Serial	Block	Circ.	Unc.
AR-16-1-1	1 cent	orange	black	red	A, B,D	25	40
AR-16-1-5	5 cents	green	black	red	B, D	35	55
AR-16-1-10	10 cents	pink	black	red	B, D	55	85
AR-16-1-25	25 cents	violet	black	red	B, D	90	150
AR-16-1-50	50 cents	peach	black	red	B, D	125	200
AR-16-1-100	1 dollar	gray	black	red	D	175	350

Monticello POW Base Camp

Located in the southeast corner of Arkansas, 90 miles southeast of Little Rock, this POW camp opened in September 1943 and held mostly Italian prisoners. When Italy capitulated, most of the prisoners were transferred to ISUs (Italian Service Units) and Monticello was used to hold Italian troublemakers, who were expelled from ISUs and high ranking officers. The camp closed in September, 1945.

AR-24-1-1

AR-24-1-5

No.	Denomination	Paper	Ink	Serial	Block	Circ.	Unc.
AR-24-1-1	1 cent	pink	black	red		150	300
AR-24-1-5	5 cents	rose	black	red		150	300

Pine Bluff Arsenal POW Camp

The War Department established the Pine Bluff Arsenal in November 1941 as a manufacturing center for magnesium and thermite munitions. It was located eight miles northwest of the Pine Bluff or 30 miles southeast of Little Rock. Between October 1944 and December 1945, 554 German POWs were incarcerated at Pine Bluff. No chits are known from this location.

Camp Joseph T. Robinson POW Base Camp

This base camp was located seven miles north of Little Rock at the site of the World War I Camp Pike, which was built in 1917. Between wars it was the headquarters of the Arkansas National Guard. With the start of World War II it took on the role of a replacement training center and the name was changed to Camp Joseph T. Robinson, in honor of a prominent Arkansas Democrat who served as an Arkansas Governor, US Senator, Senate Majority Leader, and was a candidate for Vice President. The camp was the training site for several units of Japanese Americans who volunteered for service in the US Army. The POW camp at Joseph T. Robinson opened in 1943, had peak population of 2,290, and closed in July 1946.

AR-5-1-1

AR-5-1-10

AR-5-2-1

AR-5-3-1

No.	Denomination	Paper	Ink	Serial	Block	Circ.	Unc.
AR-5-1-1	1 cent	tan	blue	red	B	150	300
AR-5-1-10	10 cent	yellow	blue	red	L	175	350
	Exchange						
AR-5-2-1	1 cent	pale green	green	red	B	75	150
	Merchandise						
AR-5-3-1	1 cent	white	blue	red	E	100	200

Rohwer Relocation Center

Rohwer Relocation Center was only 28 miles north of Jerome. The first ethnic Japanese arrived in September 1942, while the relocation camp was still in the construction phase. An additional 1,475 internees arrived before the camp was completed in January 1943. Conditions were crude in the camp, but were improved as internees adjusted to the surroundings. Some of the local population complained that the internees in the camp had much better living conditions than they had. Some examples were that the camp had electricity not available to many locals and that a train load of fruit, vegetables, and other perishable food was delivered to the camp once a week. When the Jerome camp closed on June 30, 1944, many of those interned were transferred to Rohwer. Rohwer did not close until November 30, 1945.

CALIFORNIA

	Camp	Base Camp
1	Angel Island	
2	Arbuckle	Camp Haan
3	Ashview	Fort Ord
4	Bacon Island	Fort Stockton
5	Benicia Ars.	Fort Stockton
6	Birmingham G.H.	
7	Blythe	Florence AZ
8	Boswell Ranch	Camp Cooke
9	Bouldin Island	Fort Stockton
10	California QMSD	
11	Cama	Florence AZ
12	*Camp Anza	
13	*Camp Beale	
14	*Camp Cooke	
15	*Camp Haan	
16	*Camp Roberts	
17	*Camp Stoneman	
18	Chino	Pomona
19	Clarkesburg	Fort Stockton
20	Coalinga	Fort Ord
21	Coronado AMF	
22	Dairyland	Fort Ord
23	Davis	Camp Beale
24	Delano	Camp Cooke
25	*DeWitt G. H	
26	Dos Palos	Fort Ord
27	Edwards Ranch	Camp Cooke
28	Firebaugh	Fort Ord
29	Five Points	Fort Ord
30	*Fort Ord	

31	Garden Grove	Pomona Ord. Depot
32	Goleta	Camp Cooke
33	Griffith Park	
34	King Island	Stockton
35	Kohler	Stockton
36	Irwin	Camp Haan
37	Lakelands	Camp Cooke
38	Lamont	Camp Cooke
39	Lathrop	Stockton
40	Lemoore	Camp Cooke
41	Letterman G.H.	
42	Lockett I	Camp Haan
43	Lockett II	Camp Haan
44	Los Angeles POE	
45	Madera	Fort Ord
46	Mendota	Fort Ord
47	McClellan Field	Stockton
48	Mitchell GH	
49	Napa	Camp Beale
50	Old River	Camp Cooke
51	Palm Springs	Camp Haan
52	*Pomona Ord. Depot	
53	Rankin	Camp Cooke
54	Ridge Tract	Stockton
55	Roberts Island	Stockton
56	Sacmento Sig. Dep.	Stockton
57	Santa Anna AB	Pomona Ord. Depot
58	San Fernando	Pomona Ord. Depot
59	San Joaquin	Stockton
60	San Luis Obispo	
61	Saticoy	Camp Cooke
62	Shafter	Camp Cooke

63	Shoemaker	
64	Soledad	Fort Ord
65	*Stockton	
66	Tachi Farms	Camp Cooke
67	Tagus Ranch	Camp Cooke
68	Tipton	Camp Cooke
69	Torney G.H.	
70	Tracy QMSD	
71	Tranquility	Fort Ord
72	Treasure Island	
73	Tulare Fair Grounds	Camp Cooke
74	Tulare Ragus Ranch	Camp Cooke
75	Tulelake	Camp White, OR
76	Upper Jones Tract	Stockton
77	Vernalis	Stockton
78	Windsor	Camp Beale

Anza POW Camp

CA-12-1-10

No.	Denomination	Paper	Ink	Serial	Block	Circ.	Unc.
CA-12-1-10	10 cent Dec. 31, 1946	orange	black	black	A	175	350

Camp Beale POW Base Camp

Camp Beale opened for German POWs in March 1944. In 1944, existing dormitories and storehouses at the site were converted into the POW camp. When the conversion was finished, there were 17 barracks, four mess halls, and a canteen. Profits from the canteens went to buy more than 5,000 books for the POW camp library. It was a base-camp for a group of small camps in northern California. They were located at Arbuckle, in Colusa County (200 Germans); Chico, in Butte County (475 Germans); Davis, in Yolo County (250 Germans); Napa, in Napa County (250 Germans); and Windsor, in

Sonoma County (250 Germans). The camps provided agricultural help to local farms and ranchers. The camp closed in April 1946.

CA-13-1-1

CA-13-1-5

CA-13-1-10

CA-13-1-25

CA-13-1-50

No.	Denomination	Paper	Ink	Serial	Block	Circ.	Unc.
CA-13-1-1	1 cent	yellow	black/red	black	A	50	100
CA-13-1-5	5 cents	blue	black/red	black	A	50	100
CA-13-1-10	10 cents	orange	black/red	black	A	75	150
CA-13-1-25	25 cents	pink	black/red	black	A	100	200
CA-13-1-50	50 cents	pale violet	black/red	black	A	200	400

Camp Cooke POW Base Camp

Camp Cooke located in Santa Barbara County was opened for POWs in October 1944. This base camp had 829 German POWs in residence. It was a base camp for 12 satellite camps in and around Santa Barbara County. The camp closed in May 1946.

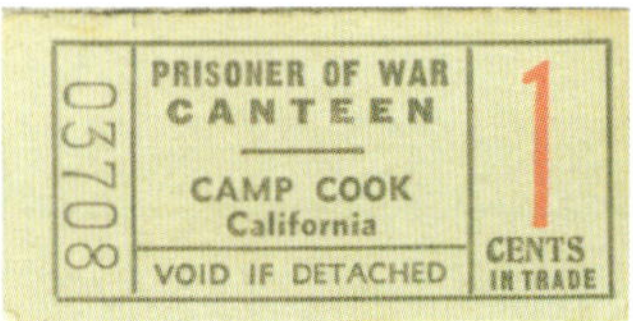

CA-14-1-1

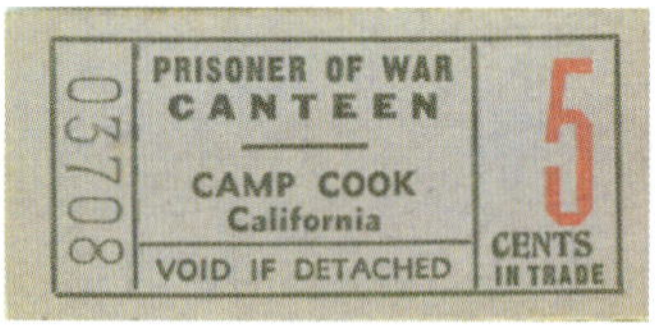

CA-14-1-5

CA-14-2-1

CA-14-2-5

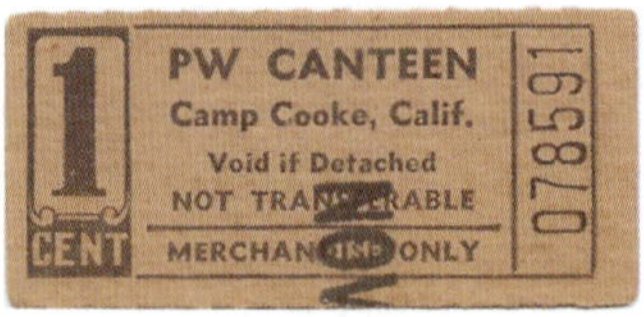

CA-14-3-1

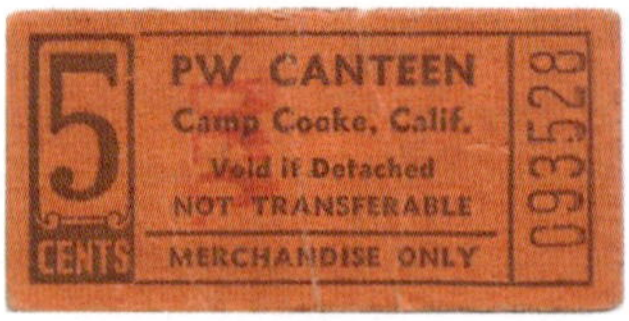

CA-14-3-5

CA-14-3-10

No.	Denomination	Paper	Ink	Serial	Block	Circ.	Unc.
	Camp Cook						
CA-14-1-1	1 cent	white	black/red	black		150	250
CA-14-1-5	5 cents	blue	black/red	black		150	250
	Camp Cooke						
CA-14-2-1	1 cent	green	black	red		150	250
CA-14-2-5	5 cents	rose	black	red		150	250
	PW Canteen Camp Cooke						
CA-14-3-1	1 cent	brown	black	black		150	250
CA-14-3-5	5 cents	red	black	black		150	250
CA-14-3-10	10 cents	tan	black	black		175	300

Camp Haan POW Base Camp

Camp Haan in Riverside County was established in 1940 from 8,058 acres of adjacent March Air Base as a training base for anti-aircraft artillery units. In October 1944, a POW camp holding German POWs and Italian Service Units was formed. A total of 600 German POWs and ISU Italian members were at the camp. The camp closed in May of 1946.

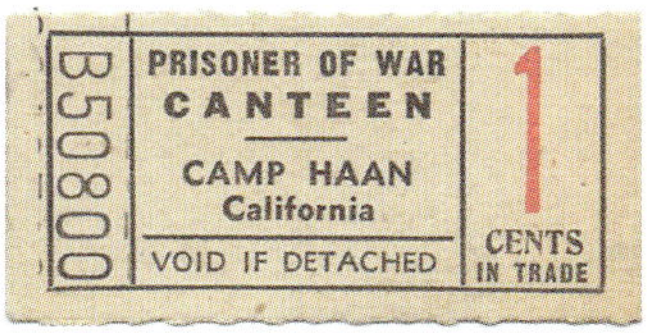

CA-15-1-1

CA-15-1-5

CA-15-1-10

No.	Denomination	Paper	Ink	Serial	Block	Circ.	Unc.
CA-15-1-1	1 cent	white	black/red	black	B	125	250
CA-15-1-5	5 cents	gray	black/red	black	B	175	350
CA-15-1-10	10 cents	pale yellow	black/red	black	B	175	350

Camp Stoneman POW Base Camp

Camp Stoneman was one of the largest troop staging areas for the Pacific. The POW camp at Stoneman opened in July 1945. The POW camp was located in Contra Costa County and closed in June 1946.

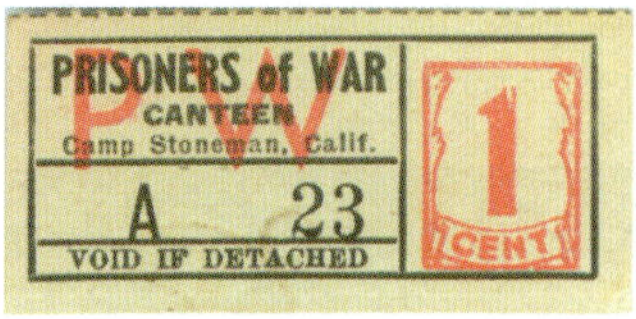

CA-17-1-1

CA-17-1-5

CA-17-1-10

CA-17-2-1a

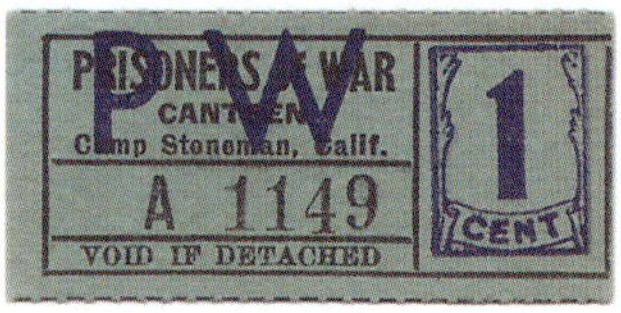

CA-17-2-1b

CA-17-2-1c

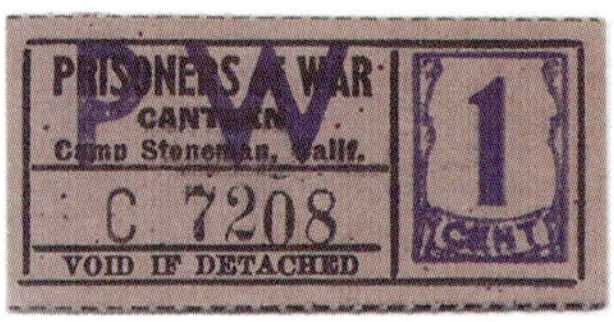

CA-17-2-1d

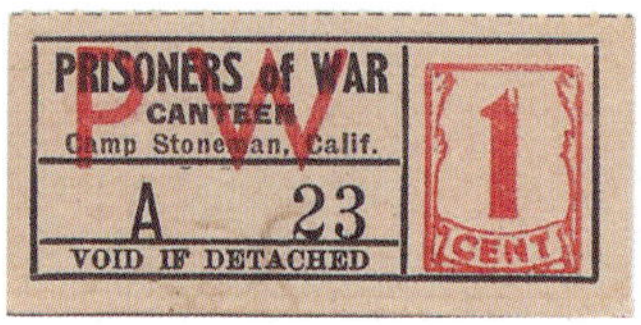

CA-17-2-1e

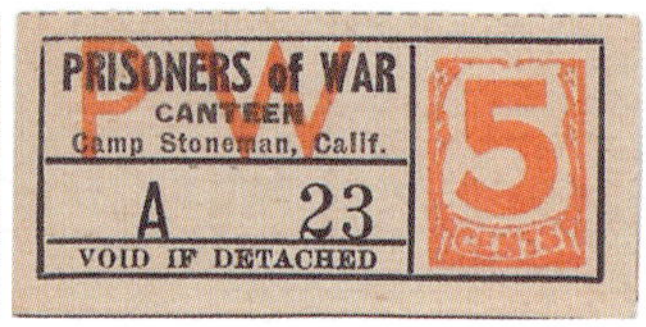

CA-17-2-5a

CA-17-2-5b

CA-17-2-5c

CA-17-2-10a

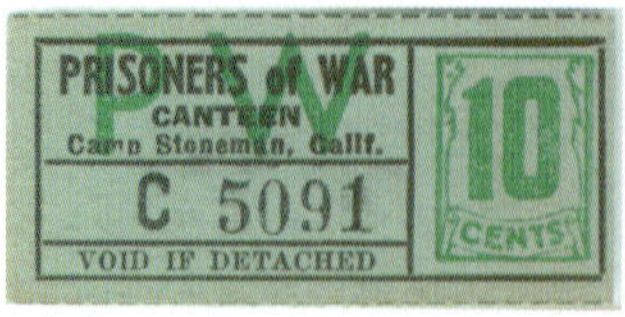

CA-17-2-10b

CA-17-2-25

No.	Denomination	Paper	Ink	Serial	Block	Circ.	Unc.
CA-17-1-1	1 cent no P of W on back	pale tan	black/red	black	A	125	250
CA-17-1-5	5 cents no P of W on back	pale tan	orange/black	black	A	125	250
CA-17-1-10	10 cents no P of W on back	pale tan	purple/black	black	A	175	350
CA-17-2-1a	1 cent	yellow	black/red	black	A,C	75	150
CA-17-2-1b	1 cent	blue	black/purple	black	A,C	75	150
CA-17-2-1c	1 cent	salmon	black/purple	black	A,C	75	150
CA-17-2-1d	1 cent	violet	black/purple	black	A,C	75	150
CA-17-2-1e	1 cent	pink	black/red	black	A,C	75	150
CA-17-2-5a	5 cents	pink	black/red	black	A,C	75	150
CA-17-2-5b	5 cents	salmon	black/red	black	A,C	75	150
CA-17-2-5c	5 cents	violet	black/red	black	A,C	75	150
CA-17-2-10a	10 cents	salmon	black/green	black	A,C	75	150
CA-17-2-10b	10 cents	green	black/green	black	A,C	75	150
CA-17-2-25	25 cents	orange	black/red	black	A,C	175	350

DeWitt General Hospital POW Base Camp

The prisoner of war camp at DeWitt General Hospital in Placer County was established June 12, 1945 for the treatment of German POWs. A total of 200 German POWs worked at the hospital. The camp closed on March 1, 1946.

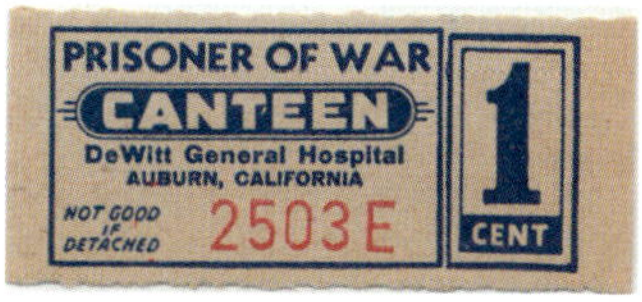

CA-25-1-1

CA-25-1-5

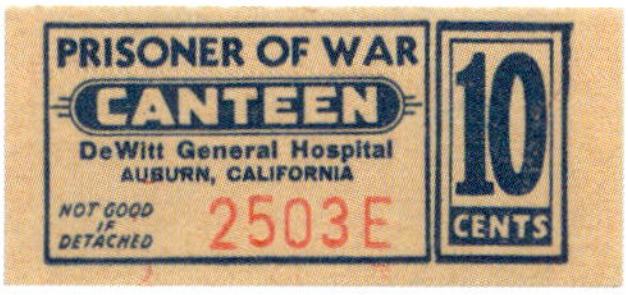

CA-25-1-10

CA-25-1-25

No.	Denomination	Paper	Ink	Serial	Block	Circ.	Unc.
CA-25-1-1	1 cent	gray	blue	red	E	25	40
CA-25-1-5	5 cents	pink	blue	red	E	35	55

CA-25-1-10	10 cents	yellow	blue	red	E	55	85
CA-25-1-25	25 cents	peach	blue	red	E	90	150

Ft. Ord POW Base Camp

Fort Ord in Monterey County was selected to hold German prisoners of war. The first German POWs arrived in March 1943, and there was an unsuccessful attempt by some 500 of them to escape by a tunnel in August 1944. The POW camp closed in June 1946.

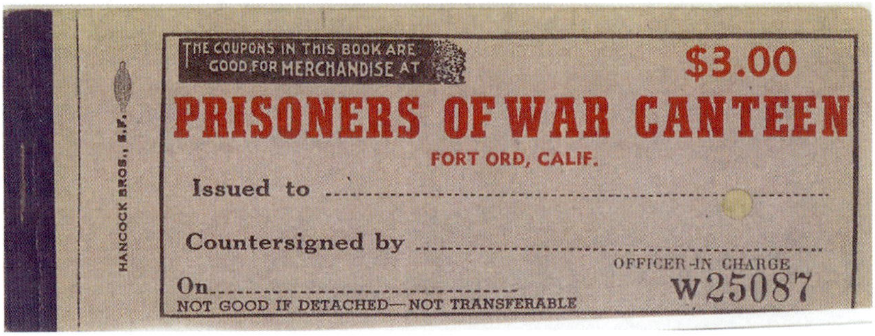

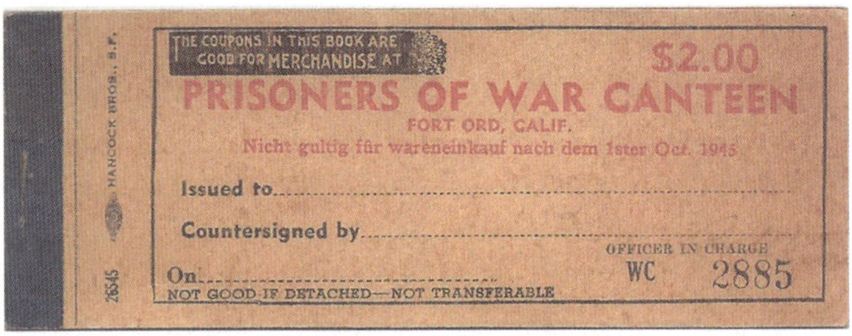

CA-30-1-1a

CA-30-1-1b

CA-30-1-5a

CA-30-1-5b

CA-30-1-10a

CA-30-1-10b

CA-30-1-25

No.	Denomination	Paper	Ink	Serial	Block	Circ.	Unc.
CA-30-1-1a	1 cent	carmine	black/red	black	W D	50	100
CA-30-1-1b	1 cent	violet	black/red	black	W	50	100
CA-30-1-5a	5 cents	violet	black/red	black	W	75	150
CA-30-1-5b	5 cents	gray	black/red	black	WC&W	75	150
CA-30-1-10a	10 cents	gray	black/red	black	W C	75	175
CA-30-1-10b	10 cents	green	black/red	black	W	75	175
CA-30-1-25	25 cents	yellow	black/red	black	W	150	300

Pomona POW Camp

CA-52-1-1

CA-52-2-1

No.	Denomination	Paper	Ink	Serial	Block	Circ.	Unc.
	Pomona						
CA-52-1-1	1 cent Sep 30 1945	green	black	black	C	150	300
	SCU 1984 OSCS						
CA-52-2-1	1 cent Sep. 30 1945	brown	black	black	A	150	300

Roberts POW Camp

CA-16-1-5

CA-16-1-10

CA-16-2-1

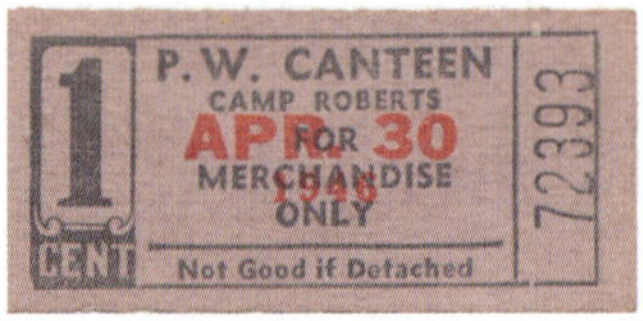

CA-16-3-1

No.	Denomination	Paper	Ink	Serial	Block	Circ.	Unc.
	31-Oct-45						
CA-16-1-5	5 cents	red	black	black		150	300
CA-16-1-10	10 cents	red	black	black		175	350
	28-Feb-46						
CA-16-2-1	1 cent	yellow	black	black		175	350
	30-Apr-46						
CA-16-3-1	1 cent	violet	black	black		175	350

Stockton POW Base Camp

The POW camp at Stockton in San Joaquin County held over 1,500 POWs from April 1944 until June 1946. It was a base camp for 15 satellite camps that held an additional 4,000 POWs, working mostly in the agricultural area. The camp closed in June 1946.

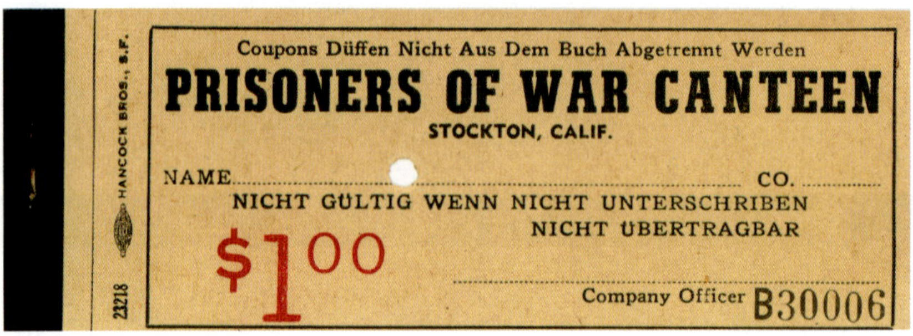

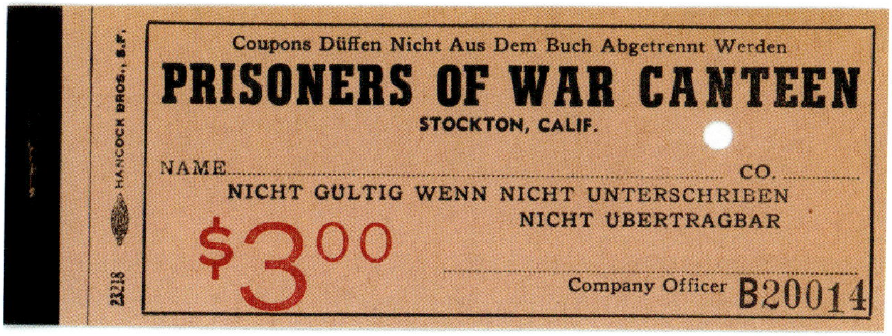

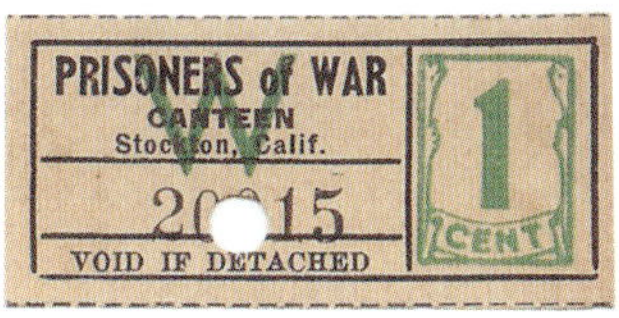

CA-65-1-1

CA-65-1-5

CA-65-1-10

CA-65-1-25

CA-65-2-1

CA-65-2-5

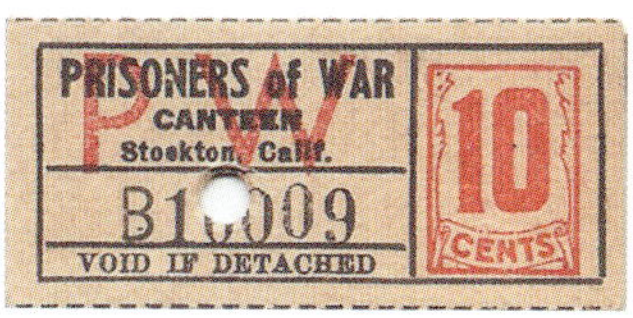

CA-65-2-10a

CA-65-2-10b

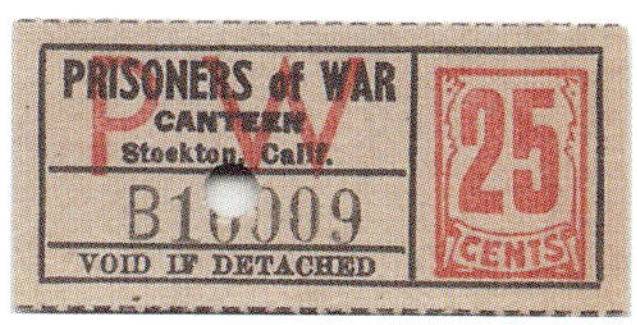

CA-65-2-25

CA-65-3-25

CA-65-4-1

No.	Denomination	Paper	Ink	Serial	Block	Circ.	Unc.
	W Overprint		green				
CA-65-1-1	1 cent	yellow	black/green	black		40	100
CA-65-1-5	5 cents	orange	black/green	black		55	125
CA-65-1-10	10 cents	white	black/green	black		75	150
CA-65-1-25	25 cents	violet	black/green	black		125	225
	PW Overprint		red				
CA-65-2-1	1 cent	yellow	black/red	black		75	150
CA-65-2-5	5 cents	gray	black/red	black	C	75	150
CA-65-2-10a	10 cents	tan	black/red	black	B	75	150
CA-65-2-10b	10 cents	gray	black/red	black	C	75	150
CA-65-2-25	25 cents	violet	black/red	black	B	125	225
	Aug 31, 1945 PW		red				
CA-65-3-25	25 cents	tan	black/red	black	A	175	350
	May 31, 1946 PW		blue				
CA-65-4-1	1 cent	blue	black/blue	black	K	175	350

COLORADO

There were three base camps located in Colorado— Camp Carson, Greeley, and Trinidad which, along with Ft. Crook, administered forty-six branch camps in the state.

	Camp	Base Camp
1	Ault	Camp Carson
2	Bob Creek	Camp Carson
3	Brighton	Camp Carson
4	Brush	Camp Carson
5	*Camp Carson	
6	Fort Collins	Camp Carson
7	Dead Man Mt.	Greeley
8	Delta	Camp Carson
9	Eaton	Camp Carson
10	Fitzsimmons G.H.	Fort Crook NE
11	Fraser	Greeley
12	Fruita	Camp Carson
13	Galeton	Camp Carson
14	Gilcrest	Camp Carson
15	Glenwood Springs	Camp Carson
16	Gould	Greeley
17	Grand Junction	Camp Carson
18	*Greeley	
19	Hale	Trinidad
20	Holly	Camp Carson
21	Johnstown	Camp Carson
22	Keenesburg	Camp Carson
23	Kersey	Camp Carson
24	Kremmling	Greeley
25	Lamar	Trinidad
26	Las Animas	Trinidad
27	Fort Logan	Camp Carson

28	Longmont	Camp Carson
29	Loveland	Camp Carson
30	Minturn	Camp Carson
31	Monte Vista	Trinidad
32	Montrose	Camp Carson
33	Fort Morgan	Camp Carson
34	New Castle	Camp Carson
35	Ovid	Camp Carson
36	Pando	Greeley
37	Pierce	Camp Carson
38	Rocky Ford	Camp Carson
39	Rocky Mountain Ars.	Camp Carson
40	Springfield	Trinidad
41	Sterling	Camp Carson
42	Stonewall (Weston)	Trinidad
43	Sugar City	Camp Carson
44	*Trinidad	
45	Walden	Greeley
46	Walsh	Camp Carson
47	Weston	Trinidad
48	Wiggins	Camp Carson
49	Yuma	Greeley

Camp Carson Base Camp

Camp Carson, located near Colorado Springs, was established in 1942 following Pearl Harbor. The POW camp opened on January 1, 1943 and held primarily Italian POWs and some German POWs. During its existence it held 9,000 POWs that did general farm work, such as canning tomatoes, cutting corn, and cutting lumber on the Western Slope. The camp closed July 1946.

CO-5-1-1a

CO-5-1-1b

CO-5-1-1c

No.	Denomination	Paper	Ink	Serial	Block	Circ.	Unc.
CO-5-1-1a	1 cent	pink	black	red	F	150	300
CO-5-1-1b	1 cent	pale orange	red	black	K	125	250
CO-5-1-1c	1 cent	pink	red	black	L	125	300

Greeley Base Camp

Greeley POW camp, also known as Camp 202, was located 50 miles northwest of Denver. Greeley held 3,000 POWs, was built in 1943, and opened in 1944. The camp was like a small town with a fire station, hospital, theater, library, and classrooms. The men had classes in English, history, and math. Some were in the camp orchestra and others sang in the choir. The soccer teams even had t-shirts in different colors made out of homemade dyes from vegetables. The camp closed in early 1946.

CO-18-1-1

CO-18-1-5

CO-18-1-10

CO-18-1-25

CO-18-2-1a

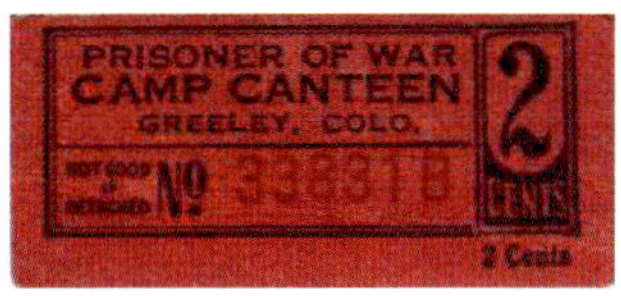

CO-18-2-2a

CO-18-2-2b

CO-18-2-5a

CO-18-2-5b

No.	Denomination	Paper	Ink	Serial	Block	Circ.	Unc.
CO-18-1-1	1 cent	orange	black	red		40	100
CO-18-1-5	5 cents	yellow	black	red		55	125
CO-18-1-10	10 cents	gray	black	red		85	150
CO-18-1-25	25 cents	pink	black	red		125	250
	Camp Canteen						
CO-18-2-1a	1 cent (D stamp)	blue	blue	red	A	100	200
CO-18-2-2a	2 cents	red	black	red	B	125	250

CO-18-2-2b	2 cents	red	black	red	C	125	250
CO-18-2-2c	2 cents (blue D stamp)	red	black	red	D	125	250
CO-18-2-2b	2 cents (red J stamp)	rose	blue	red	A	125	250
CO-18-2-5a	5 cents	yellow	blue	red	C	150	300
CO-18-2-5b	5 cents (red J stamp)	pink	blue	red	D	150	300
CO-18-2-5c	5 cents (red J stamp)	pale yellow	blue	red	A	150	300
CO-18-2-10	10 cents	pale pink	blue	red	C	175	350

Trinidad POW Base Camp

Trinidad is located on I-25, 200 miles south of Denver and 15 miles north of the New Mexico border. The 1.8 million dollar camp was opened on June 4, 1943 and housed 3,500 German POWs, most of whom were elite members of Rommel's Afrika Corps. The camp had 330 structures, including 10 guard towers and its own sewage and water systems. Courses were made available to the POWs in classes offered by the University of Colorado and the University of Denver. Shortly after the camp opened, three POWs that escaped were shot and killed. The camp was closed in 1946.

CO-44-1-1

No.	Denomination	Paper	Ink	Serial	Block	Circ.	Unc.
CO-44-1-1	1 cent	pink	black/red	red	A	150	300

CONNECTICUT

Connecticut did not have any base camps; only a branch camp, Bradley Field, that was administered by the Devens, MA base camp.

DELAWARE

Ft. DuPont and Ft. Miles were two base camps located in Delaware. Ft. DuPont administered nine branch camp locations, while Ft. Miles had no branches.

	Camp	Base Camp
1	Bethany Beach	Fort DuPont
2	Bridgeville	Fort DuPont
3	*Fort DuPont	
4	Georgetown I	Fort DuPont
5	Georgetown II	Fort DuPont
6	Harrington	Fort DuPont
7	Leipsic	Fort DuPont
8	Lewes	Fort DuPont
9	Fort Miles	
10	New Castle	Fort DuPont
11	Fort Saulsbury	Fort DuPont

Ft. DuPont POW Base Camp

Ft. DuPont, named after Rear Admiral Samuel Francis DuPont and established in 1863, is located in Delaware City. In WWII it was used as a mobilization station for deploying units overseas. In May 1944, the 1231st SCU POW camp was established using temporary buildings. Approximately 3,000 German POWs were housed in the camp. Some of the POWs were from the U-858 submarine that surrendered off the coast of Lewes, Delaware. POWs worked as dishwashers, grocers, butchers, and in other support roles for local companies. The camp closed in November 1945.

DE-3-1-1

DE-3-1-10a

DE-3-1-10b

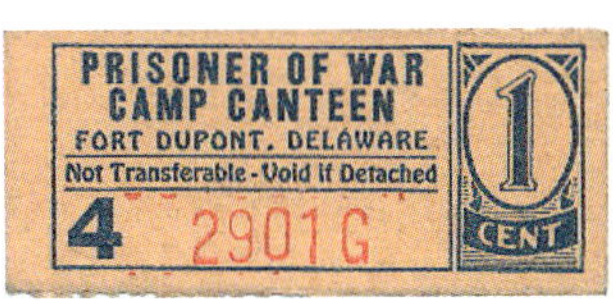

DE-3-2-1a

DE-3-2-1b

DE-3-2-5a

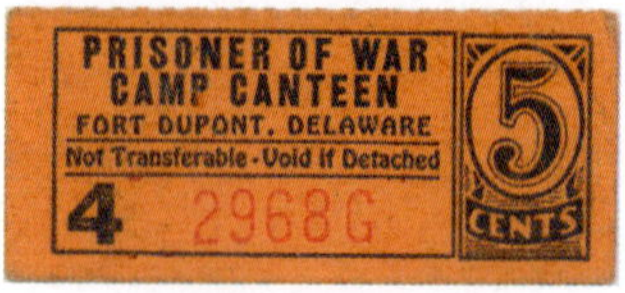

DE-3-2-5b

No.	Denomination	Paper	Ink	Serial	Block	Circ.	Unc.
	P O W Canteen						
DE-3-1-1	1 cent	pink	blue	red	C	75	150
DE-3-1-5	5 cents		blue	red		125	200
DE-3-1-10a	10 cents	green	red	red	C	150	225
DE-3-1-10b	10 cents	rose	blue	red	C	150	225
DE-3-1-25	25 cents					175	250
DE-3-1-50	50 cents					225	300
	Camp Canteen						
DE-3-2-1a	1 cent (4)	yellow	blue	red	G	75	150
DE-3-2-1b	1 cent (5)	pink	blue	red	H	125	200
DE-3-2-5a	5 cents (2)	salmon	blue	red	F	125	200
DE-3-2-5b	5 cents (4)	orange	blue	red	G	125	200
DE-3-2-10	10 cents					150	225
DE-3-2-25	25 cents					175	250
DE-3-2-50	50 cents					225	300

Ft. Miles POW Base Camp

Ft. Miles was located in Sussex County. Little is known about the camp other than it was opened November 1945, closed March 1946, and held 1,500 German POWs.

FLORIDA

	Camp	Base Camp
1	Banana River	Camp Blanding
2	Belle Glade	Camp Blanding
3	Belle Haven	Camp Blanding
4	*Camp Blanding	
5	*Camp Gordon Johnston	
6	Clewiston	Camp Blanding
7	Dade City	Camp Blanding
8	Drew Field	Camp Blanding
9	Eglin Field	Camp Blanding
10	Green Cove Springs	Camp Blanding
11	Hastings	Camp Blanding
12	Homestead AAF	Camp Blanding
13	Jacksonville NAS	Camp Blanding
14	Kendall	Camp Blanding
15	Leesburg	Camp Blanding
16	Dale Mabry Field	Camp Gordon Johnston
17	MacDill Field	Camp Blanding
18	Melbourne	Camp Blanding
19	Marianna	Fort Benning GA
20	Orlando AAB	Camp Blanding
21	Page Field	Camp Blanding
22	Telogin	Camp Gordon Johnston
23	Venice AAF	Camp Blanding
24	Welch G.H.	Camp Blanding
25	Whiting Field	Camp Rucker AL
26	White Springs	Camp Blanding
27	Winter Haven	Camp Blanding

Camp Blanding POW Base Camp

Camp Blanding was located 10 miles east of Starke, which is 35 miles southwest of Jacksonville. The camp was named after Lt. General Albert H. Blanding, a Florida native who distinguished himself during Mexican Border Service and in World War I. Camp Blanding was one of the first camps in the United States to receive foreign captives. During the summer of 1942, a group of enemy aliens from Latin America were interned at Camp Blanding. This group was quickly moved to more permanent locations in Texas, Oklahoma, and North Carolina. The enemy aliens were replaced by the first German military captives in late September. Fourteen prisoners captured from German U-boats were incarcerated at Camp Blanding, with that number quickly growing to over 200. Thus, Blanding became one of four POW camps that imprisoned German naval personnel. In November of 1943, as German Army captives arrived in the United States, a peak population of 1,243 was housed at Camp Blanding in a separate compound about a half mile from the sailors. The Blanding POW Camp closed in June of 1946. The staff at Camp Blanding also managed twenty-one sub-camps around the state of Florida.

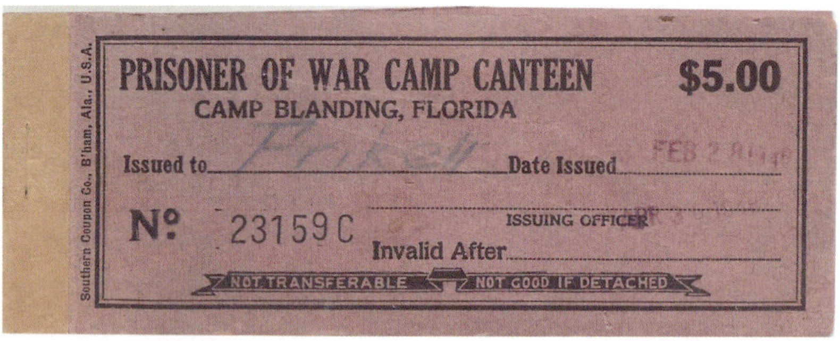

FL-4-1-5

FL-4-2-1a

FL-4-2-1b

FL-4-2-5a

FL-4-2-5b

FL-4-2-5c

FL-4-2-5d

FL-4-2-10

FL-4-2-25

No.	Denomination	Paper	Ink	Serial	Block	Circ.	Unc.
	PW Canteen						
FL-4-1-5	5 cents	rose	black	red		35	55
	Camp Canteen						
FL-4-2-1a	1 cent	white	blue	red	C,F	50	100
FL-4-2-1b	1 cent	purple	red	red	C	50	100
FL-4-2-5a	5 cents	white	blue	red	C	100	200
FL-4-2-5b	5 cents	yellow	blue	red	F	100	200
FL-4-2-5c	5 cents	orange	red	red	C	100	200
FL-4-2-5d	5 cents	rose	blue	red	C	100	200
FL-4-2-10	10 cents	orange	blue	red	C	150	300
FL-4-2-25	25 cents	white	blue	red	C	175	350

Camp Gordon Johnston POW Camp

This camp was an amphibious training center during WWII, using the Big Bend of Florida coastline for this rigorous training. The camp was originally named Camp Carrabelle (for the nearby town of that name) but changed to posthumously honor Medal of Honor winner Col. Gordon Johnston in January 1943. Col. Johnston earned his Medal of Honor for bravery during the Philippine Insurrection of 1906. Camp Gordon Johnston covered 165,000 acres of the Florida Panhandle and is 55 miles southwest of Tallahassee. The POW camp opened in March of 1944, housed a maximum of 587 prisoners, and closed in March of 1946. The prisoners performed work on the military installation. Chits were just recently discovered from this camp.

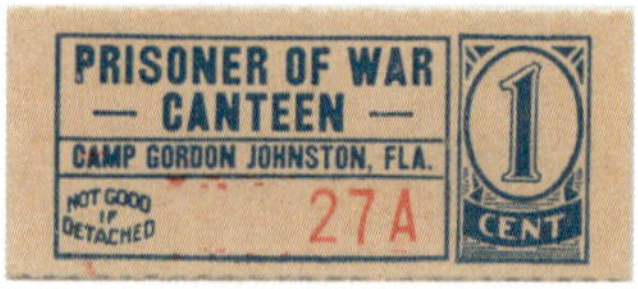

FL-5-1-1

No.	Denomination	Paper	Ink	Serial	Block	Circ.	Unc.
FL-5-1-1	1 cent	white	blue	red	A	175	350

GEORGIA

Georgia had POW camps in both World War I and II. During World War II there were 37 camps with base camps at Ft. Benning and Camp Wheeler.

	Camp	Base Camp
1	Albany	Opelika AL
2	Americus	Fort Benning
3	Ashburn	Camp Wheeler
4	Atlanta Ord. Dep.	Fort Benning
5	Axton	Fort Benning
6	Bainbridge AAF	Fort Benning
7	Battey G.H.	Fort Oglethorpe
8	*Camp Wheeler	
9	Chatham AAF	Stewart
10	Daniel Field	Camp Wheeler
11	Dublin	Camp Wheeler
12	Fargo	Fort Benning
13	Finney G.H.	Fort Benning
14	*Fort Benning	
15	Fort Oglethorpe	
16	Glynco NAS	Fort Benning
17	Gordon	
18	Griffin	Camp Wheeler
19	Hunter Field	Stewart
20	Jessup	Camp Wheeler
21	Monticello	Camp Wheeler

22	Moody Field	Fort Benning
23	Moultrie	Fort McClellan AL
24	Perry	Camp Wheeler
25	Reidsville	Stewart
26	Savannah ASF	Stewart
27	Spence Field	Fort Benning
28	Statesboro	Gordon
29	Stewart	
30	Tifton	
31	Toccoa	Croft SC
32	Turner Field	Fort Benning
33	Valdosta	Opelika AL
34	Wadely	Gordon
35	Waycross	Fort Benning
36	Waynesboro	Camp Wheeler

Ft. Benning POW Base Camp

Fort Benning is located in Columbus, at the border of Alabama and Georgia in Chattahoochee County. The fort was established in 1917 as a school to produce combat infantry men for the US Army. The fort trained soldiers for both World War I and II. In 1942, a POW camp was established and held 2,234 POWs at its peak. The camp closed July 1946.

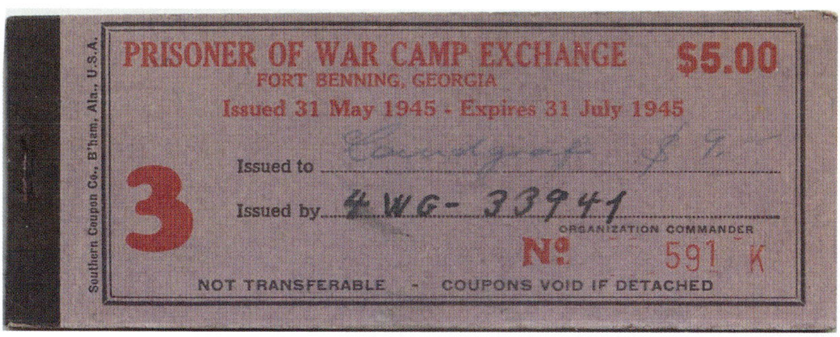

GA-14-1-1

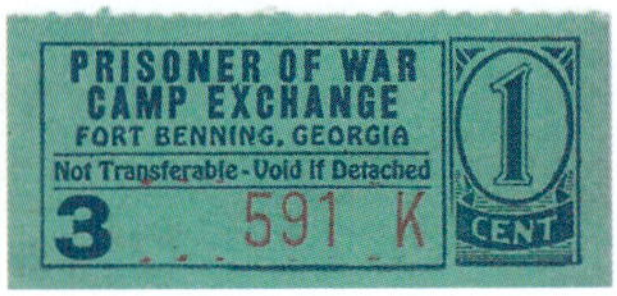

GA-14-2-1a

GA-14-2-1b

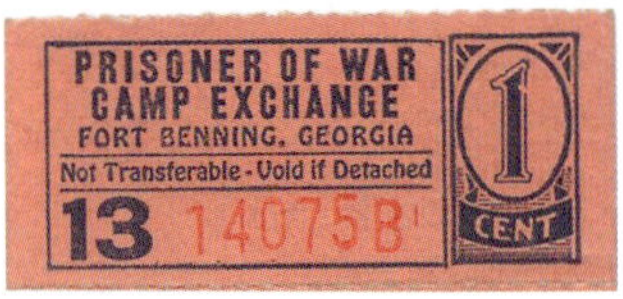

GA-14-2-1c

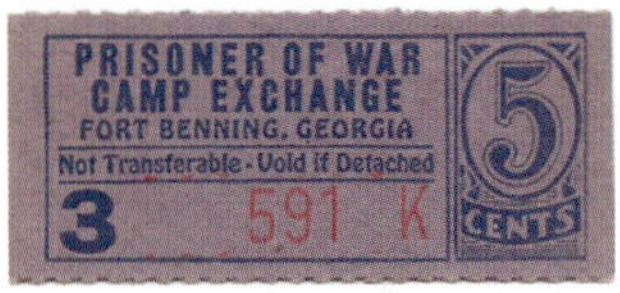

GA-14-2-5a

GA-14-2-5b

GA-14-2-10a

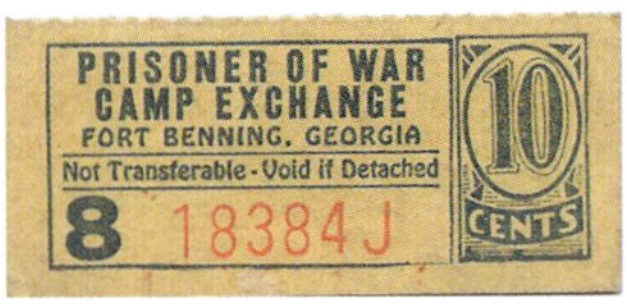

GA-14-2-10b

GA-14-2-10c

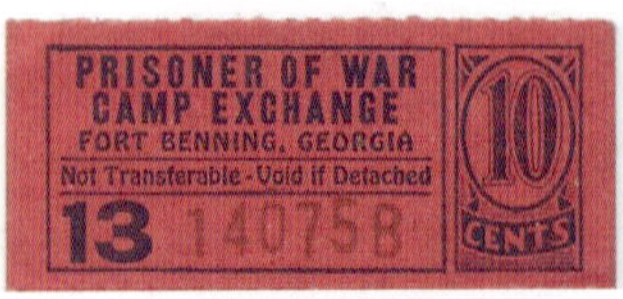

GA-14-2-10d

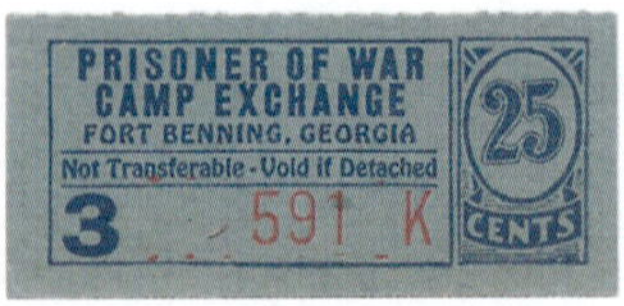

GA-14-2-25a

GA-14-2-25b

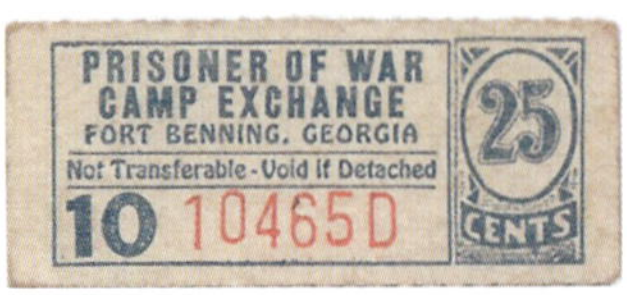

GA-14-2-25c

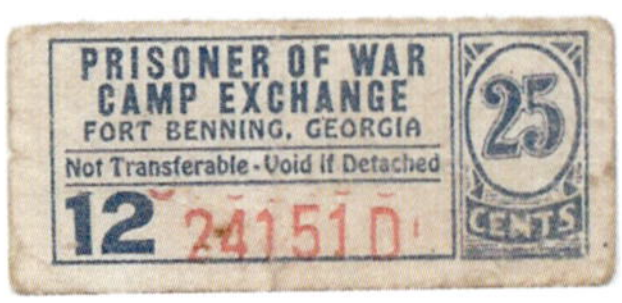

GA-14-2-25d

GA-14-3-5

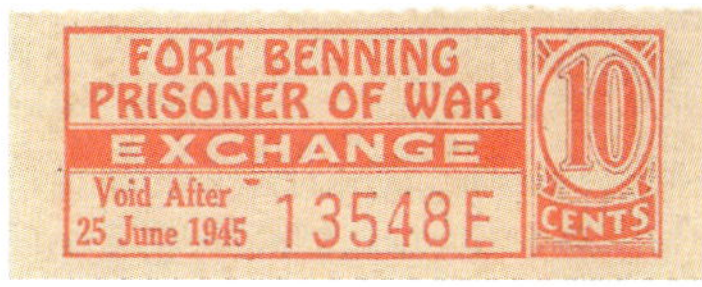

GA-14-3-10

No.	Denomination	Paper	Ink	Serial	Block	Circ.	Unc.
	Prisoner of War Exchange						
GA-14-1-1	1 cent	purple	blue	red	B	175	350
	Camp Exchange						
GA-14-2-1a	1 cent (3)	blue	blue	red	A	50	100
GA-14-2-1b	1 cent (10)	peach	blue	red	D	50	100
GA-14-2-1c	1 cent (13)	peach	blue	red	B	50	100
GA-14-2-5a	5 cents (3)	purple	blue	red	K	75	125
GA-14-2-5b	5 cents (4)	purple	blue	red	L	75	150
GA-14-2-10a	10 cents (3)	white	blue	red	K	100	200
GA-14-2-10b	10 cents (8)	yellow	blue	red	J	100	200
GA-14-2-10c	10 cents (11)	red	blue	red	D	100	200
GA-14-2-10d	10 cents (13)	red	blue	red	B	100	200
GA-14-2-25a	25 cents (3)	blue	blue	red	K	175	250
GA-14-2-25b	25 cents (8)	blue	blue	red	K	175	250
GA-14-2-25c	25 cents (10)	white	blue	red	D	175	250
GA-14-2-25d	25 cents (12)	white	blue	red	D	175	250
	25-Jun-45						
GA-14-3-5	5 cents	rose	red	red	E	75	125
GA-14-3-10	10 cents	white	red	red	E	100	175

Camp Wheeler POW Base Camp

Camp Wheeler was located near Macon in Bibb County, Georgia. The camp opened for POWs in April 13, 1943 for the first contingent of Italian POWs. With the massive influx of German POWs from the North African campaign, in May 12, 1944 the camp was switched from Italian to a German POW camp. By the end of 1944 the number of German POWs was more than 1,800. POWs worked in the camp as mechanics, typewriter experts, and tailors. The camp closed on January 19, 1946.

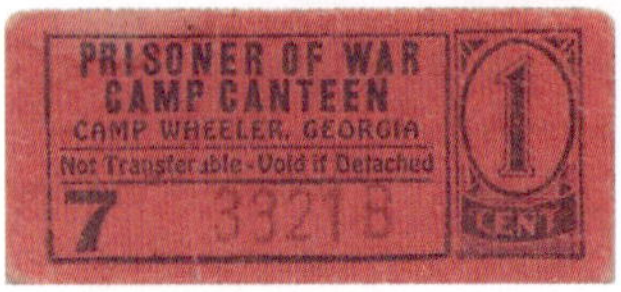

GA-8-1-1

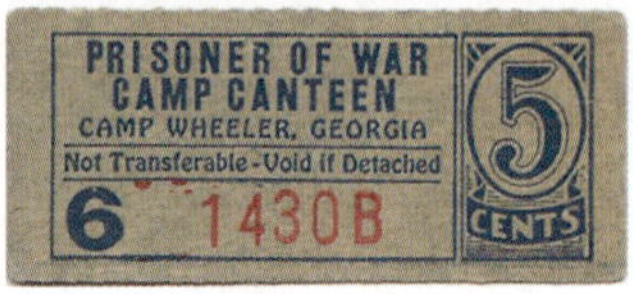

GA-8-1-5

GA-8-2-1

GA-8-2-5

GA-8-2-10

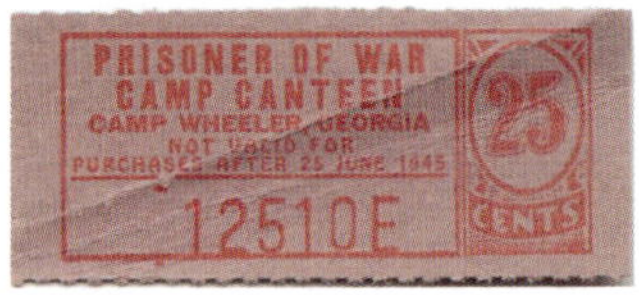

GA-8-2-25

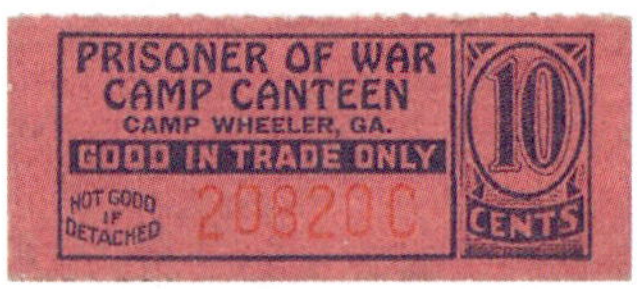

GA-8-3-10

No.	Denomination	Paper	Ink	Serial	Block	Circ.	Unc.
GA-8-1-1	1 cent	rose	blue	red	5 B	75	150
GA-8-1-1a	1 cent	rose	blue	red	7 B	75	150
GA-8-1-5	5 cent	blue	blue	red	6 B	100	200
	25 June 1945						
GA-8-2-1	1 cent	blue	red	red	D, E, F	40	90
GA-8-2-5	5 cents	yellow	red	red	E	75	125
GA-8-2-10	10 cents	orange	red	red	E	100	175
GA-8-2-25	25 cents	purple	red	red	E	200	325
	Good in Trade Only						
GA-8-3-10	10 cents	rose	blue	red	C	175	350

HAWAII

The Central Pacific POW camp was located on Oahu. The POW camp held more than 4,000 soldiers, mostly Okinawans conscripted into the Japanese army. The first POWs in the camp were conscripted Koreans captured on Gilbert Island in November 1943. Later, both Pacific and European theater POWs were held. The camp was located in Honouliuli near the town of Waipahu. Also held were 300 internees, the majority of whom were American citizens. The camp is part of the new Honoliuli National Monument designated on February 19, 2015.

	Hawaii
1	*Central Pacific

HI-1-1-1

HI-1-1-5

HI-1-1-10

No.	Denomination	Paper	Ink	Serial	Block	Circ.	Unc.
HI-1-1-1	1 cent	tan	blue	red	B	75	150
HI-1-1-5	5 cents	blue	blue	red	B	100	175
HI-1-1-10	10 cents	rose	blue	red	B	125	225

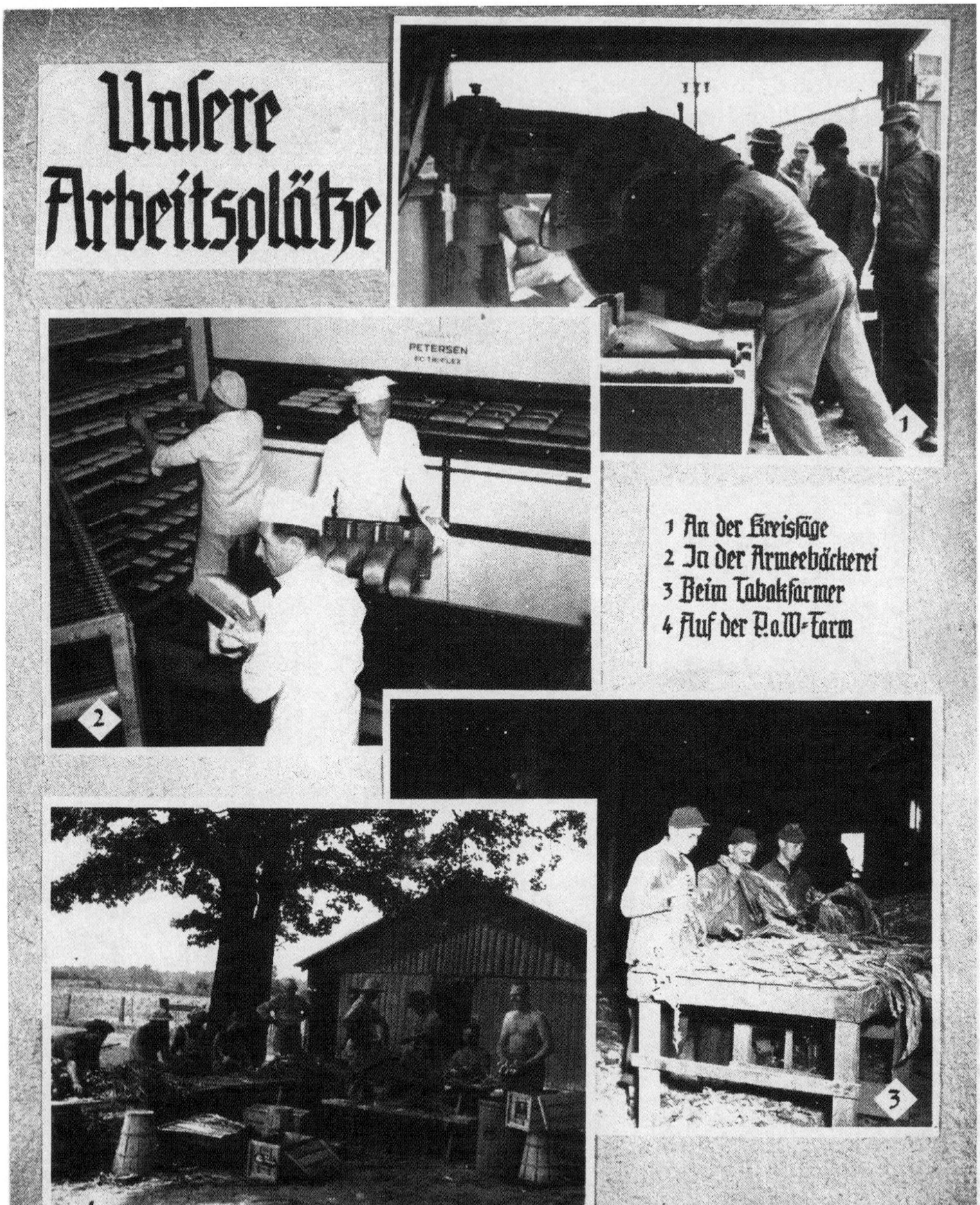

Types of Work Performed by POWs

Our Workplaces

1. At the circular saw

2. In the army bakery

3. At the tobacco farmer

4. On the POW farm

IDAHO

Idaho had two POW base camps—Camp Farragut, the base for the northern camps in Idaho and Camp Rupert, the base for the camps in the southern part of the state.

	Camp	Base Camp
1	Aberdeen	Rupert
2	Blackfoot	Rupert
3	*Farragut	
4	Filer	Rupert
5	Fort Hall	Rupert
6	Franklin	Rupert
7	Idaho Falls	Rupert
8	Nampa	Rupert
9	Marsing	Rupert
10	Pocatello	Rupert
11	Preston	Rupert
12	Rigby	Rupert
13	*Rupert	
14	Shelley	Rupert
15	Sugar City	Rupert
16	Thomas, Blackfoot	Rupert
17	Upper Deer Flat	Rupert
18	Wilder	Rupert

Camp Farragut POW Base Camp

Camp Farragut was located 30 miles north of Coeur d'Alene in Kootenai county. The POW camp opened in February of 1945 and was populated by German POWs captured in June 1944. The 748 POWs worked as gardeners and in camp maintenance. In December 1945, the commanding officer of Camp Farragut used the camp paper to send his prisoners this message in both German and English: "On your last Christmas as a Prisoner of War in America we would give you these presents to carry back to the homeland: A little view of the democratic life; A new courage to face the problems of tomorrow; A memory of a peaceful Christmas where you had Freedom of Speech, Freedom of Religion, Freedom from Want, and Freedom from Fear." The camp closed in May 1946.

ID-3-1-1

No.	Denomination	Paper	Ink	Serial	Block	Circ.	Unc.
ID-3-1-1	1 cent	gray	blue	red		175	350

Camp Rupert POW Base Camp

Camp Rupert was located in Paul, Idaho in Gem County. This base camp opened for POWs in October 1944. The camp housed 2,768 German POWs. The sub-camps that Camp Rupert supported housed an additional 12,232 POWs, who worked mostly to harvest sugar beets. The camp was noted for caring for the hardcore Nazi sympathizers, captured POWs that had escaped other camps, POWs that had staged strikes of any kind, and general malcontents. The camp closed in April 1946.

ID-13-1-1

ID-13-1-5a

ID-13-1-5b

ID-13-1-10

ID-13-1-25

ID-13-2-1

No.	Denomination	Paper	Ink	Serial	Block	Circ.	Unc.
	Camp Canteen						
ID-13-1-1	1 cent	white	blue	red		35	50
ID-13-1-5a	5 cents	blue	blue	red		45	75

ID-13-1-5b	5 cents	tan	blue	red		100	200
ID-13-1-10	10 cents	pink	blue	red		65	100
ID-13-1-25	25 cents	yellow	blue	red		125	200
	Camp						
ID-13-2-1	1 cent	white	red/black	none		175	350

ILLINOIS

There were 20 POW camps in Illinois with 3 base camps located at Camp Ellis, Camp Grant, and Ft. Sheridan.

	Camp	Base Camp
1	Arlington Heights	Fort Sheridan
2	*Camp Ellis	
3	*Camp Grant	
4	Des Plaines	Fort Sheridan
5	Eureka	Camp Ellis
6	*Fort Sheridan	
7	Gibson City	Camp Ellis
8	Hampshire	Camp Grant
9	Hoopeston	Camp Ellis
10	Joliet	Fort Sheridan
11	Lanark	Camp Grant
12	Mayo G.H	Camp Ellis
13	Milford	Camp Ellis
14	Pine	Fort Sheridan
15	Pomona	Camp Ellis
16	Skokie Valley	Fort Sheridan
17	Sycamore	Camp Grant
18	Thornton	Fort Sheridan
19	Washington	Camp Ellis
20	Waterloo	Camp Ellis

Camp Ellis Base Camp

Camp Ellis was located in Fulton County and trained army service forces. Construction started on the POW camp in September 1942. The camp opened in April 1943 with the official dedication on July 14, 1943. The first POWs were a group of 1,000 Germans that arrived in August 1943. At its peak, the camp held almost 5,000 POWs. The camp closed in March 1946.

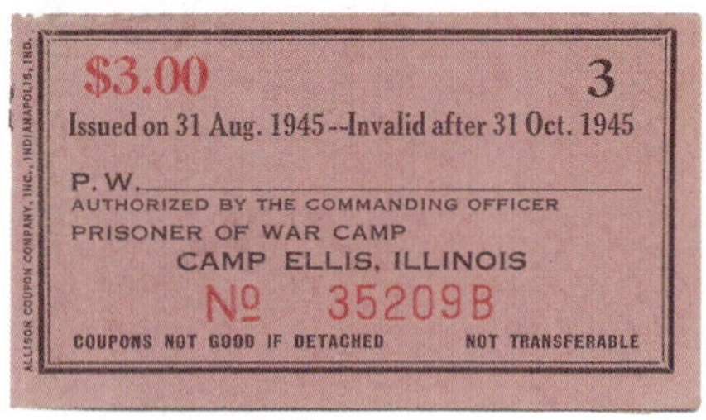

IL-2-1-1a

IL-2-1-1b

IL-2-1-1c

IL-2-1-2a

IL-2-1-2b

IL-2-1-5a

IL-2-1-5b

IL-2-1-10a

IL-2-1-10b

IL-2-1-25

IL-2-1-50

IL-2-2-1

No.	Denomination	Paper	Ink	Serial	Block	Circ.	Unc.
IL-2-1-1a	1 cent	pink	black	red		50	125
IL-2-1-1b	1 cent	pale yellow	blue	red	A, B, C	25	40
IL-2-1-1c	1 cent	pink	black	black	C	25	40
IL-2-1-2a	2 cents	pink	blue	red	C	50	100
IL-2-1-2b	2 cents	orange	black	black	C	50	100
IL-2-1-5a	5 cents	blue	blue	red	B, C	35	55
IL-2-1-5b	5 cents	purple	black	black	B	35	55
IL-2-1-5c	5 cents	green	blue	red	A	35	55
IL-2-1-10a	10 cents	brown	blue	red	B	55	85
IL-2-1-10b	10 cents	tan	black	black	B, C	55	85
IL-2-1-10c	10 cents	buff	black	red		75	150
IL-2-1-25	25 cents	red	black	black	D	90	150
IL-2-1-50	50 cents	blue	black	black	D	125	200
	Large Number On Left						
IL-2-2-1	1 cent	pink	black	red	3 B	125	250

Camp Grant POW Base Camp

Camp Grant was located on the outskirts of Rockford, Illinois and was named for Civil War General Ulysses S. Grant. During World War II it was used by the Army Medical Service.

The Camp opened for POWs in August 1943. A total of 1,669 POW were held in the camp until May of 1946, when the last of the POWs left the camp.

IL-3-1-1a

IL-3-1-1b

IL-3-1-1c

IL-3-1-2a

IL-3-1-2b

IL-3-1-5a

IL-3-1-5b

IL-3-1-10

IL-3-1-25a

IL-3-1-25b

No.	Denomination	Paper	Ink	Serial	Block	Circ.	Unc.
IL-3-1-1a	1 cent	yellow	blue	red	A	25	40
IL-3-1-1b	1 cent	gray	black	red	A	50	100
IL-3-1-1c	1 cent	pink	green	red	E	35	75
IL-3-1-2a	2 cents	white	black	red	E	50	100
IL-3-1-2b	2 cents	pink	green	red	E	50	100
IL-3-1-5a	5 cents	blue	blue	red	C, E	35	55
IL-3-1-5b	5 cents	white	black	red	E	50	100
IL-3-1-10	10 cents	yellow	blue	red	E	75	150
IL-3-1-25a	25 cents	white	black	red	E	90	175
IL-3-1-25b	25 cents	rose	blue	red	E	90	175

Ft. Sheridan POW Base Camp

Fort Sheridan was the largest of the POW camps in Illinois, located in Lake County just north of Chicago. The fort opened for POWs in October 1944. As a base POW camp, in 1944, Fort Sheridan had total administrative control of all POW camps located in the surrounding states of Illinois, Michigan, and Wisconsin. The total number of POWs was 15,000 prisoners. The Fort Sheridan camp held a total of 1,569 POWs and issued the only known 2 dollar chit. The POW camp closed in May, 1946.

IL-6-1-1a

IL-6-1-1b

IL-6-1-1c

IL-6-1-1d

IL-6-1-1e

IL-6-1-5a

IL-6-1-5b

IL-6-1-10a

IL-6-1-10b

IL-6-1-25a

IL-6-1-25b

IL-6-1-25c

IL-6-1-50

IL-6-1-1.00

IL-6-1-2.00

No.	Denomination	Paper	Ink	Serial	Block	Circ.	Unc.
IL-6-1-1a	1 cent	tan	red	red	A	25	40
IL-6-1-1b	1 cent	rose	red	red	B, H	25	40
IL-6-1-1c	1 cent	pink	black	black	C	50	100
IL-6-1-1d	1 cent	red	blue	red	F	25	40
IL-6-1-1e	1 cent	violet	black	red	H	50	100
IL-6-1-5a	5 cents	yellow	red	red	A, B, C	35	55
IL-6-1-5b	5 cents	tan	black	red	H	35	55
IL-6-1-10a	10 cents	yellow	red	red	B	55	85
IL-6-1-10b	10 cents	tan	red	red	B, C	55	85
IL-6-1-25a	25 cents	blue	red	red	C	90	150
IL-6-1-25b	25 cents	blue	blue	red	H	90	150
IL-6-1-25c	25 cents	tan	red	red	J	90	150

IL-6-1-50	50 cents	rose	red	red	J	200	400
IL-6-1-100	1 dollars	tan	red	red	J	300	600
IL-6-1-200	2 dollars	tan	red	red	J	500	1000

INDIANA

Indiana only had 10 POW camps with 2 base camps, one at Atterbury and the other at Ft. Benjamin Harrison.

	Camp	Base Camp
1	Austin	Camp Atterbury
2	*Camp Atterbury	
3	Eaton	Camp Atterbury
4	*Ft. Benjamin Harrison	
5	Indiana Ord. Works	Fort Knox KY
6	Jeffersonville	Fort Knox KY
7	Morristown	Camp Atterbury
8	Scott (Ft. Wayne)	Camp Perry OH
9	Vincennes	Camp Atterbury
10	Windfall	Camp Atterbury

Camp Atterbury POW Base Camp

Camp Atterbury was located in Johnson, Bartholomew, and Shelby counties. The first POWs to arrive were a contingent of Italians on April 30, 1943. They were transferred in January 1944 to Fort Benjamin Harrison, so that the German POWs that arrived on May 8, 1944 would not have any issues with their former ally. There were three compounds that could hold 1,000 POWs each. By "hot bunking," due to night jobs for the POWs, the camp accommodated up to 4,429 POWs. The POW camp closed in May 1946.

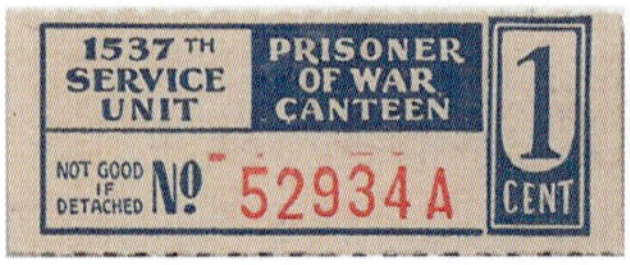

IN-2-1-1

IN-2-1-5

IN-2-1-10

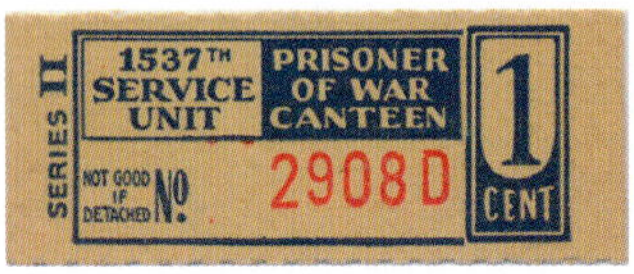

IN-2-2-1

IN-2-2-5

IN-2-2-10

IN-2-2-25

IN-2-3-1

IN-2-3-5

IN-2-3-10

IN-2-3-25

No.	Denomination	Paper	Ink	Serial	Block	Circ.	Unc.
	1537th Service Unit						
IN-2-1-1	1 cent	white	blue	red	A	25	40
IN-2-1-5	5 cents	blue	blue	red	A	35	55
IN-2-1-10	10 cents	rose	blue	red	A	55	85
	1537th Series II						
IN-2-2-1	1 cent	yellow	blue	red	D	25	40
IN-2-2-5	5 cents	tan	blue	red	D	35	55
IN-2-2-10	10 cents	yellow	blue	red	D	55	85

IN-2-2-25	25 cents	red	blue	red	D	90	150
	1560 S.C.U.						
IN-2-3-1	1 cent	pink	blue	red	C	25	40
IN-2-3-5	5 cents	red	blue	black	C	35	55
IN-2-3-10	10 cents	violet	blue	black	C	100	175
IN-2-3-25	25 cents	pale tan	blue	red	C	125	225

Ft. Benjamin Harrison POW Base Camp

Fort Benjamin Harrison was located in Indianapolis in Marion County. The camp opened for Italian POWs captured in North Africa in late April 1943. In January 1944, 250 Italian former POWs, now co-belligerents, were transferred from Camp Atterbury. In May 1944, 300 German POWs, formerly of the Afrika Corps, arrived in the POW camp. The POW camp closed in March 1945.

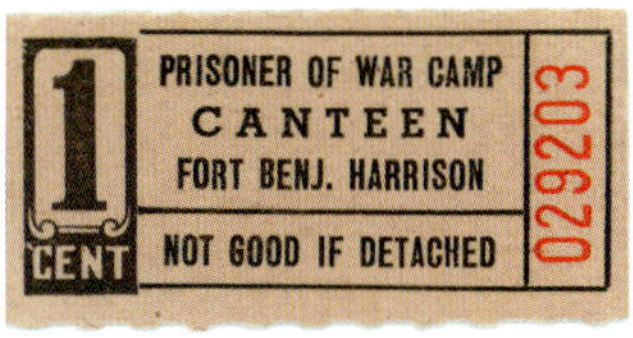

IN-4-1-1

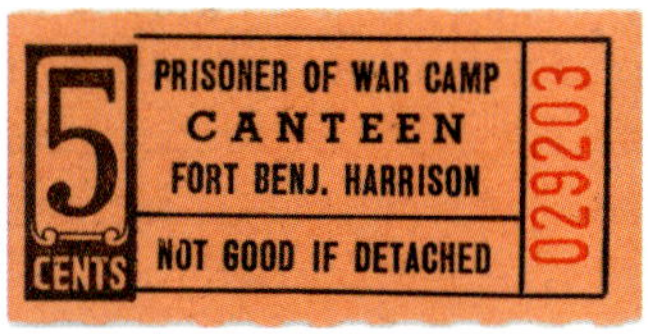

IN-4-1-5

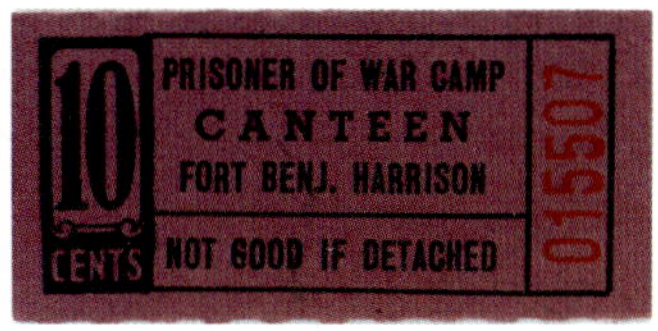

IN-4-1-10

No.	Denomination	Paper	Ink	Serial	Block	Circ.	Unc.
IN-4-1-1	1 cent	tan	black	red		100	200
IN-4-1-5	5 cents	orange	black	red		150	300
IN-4-1-10	10 cents	brown	black	red		175	350

IOWA

Iowa had two base camps that ran 12 branch camps in Iowa, as well as 17 in Minnesota and 2 in South Dakota.

	Camp	Base Camp
1	*Algona	
2	Audubon	Calrinda
3	Charles City	Algona
4	*Clarinda	
5	Clinton	Algona
6	Eldora	Algona

7	Muscatine	Algona
8	Onawa	Algona
9	Schick GH	Weingarten MO
10	Shenandoah	Algona
11	Storm Lake	Algona
12	Tabor	Algona
13	Toleoda	Algona
14	Wapello	Calrinda
15	Waverly	Algona

Algona

Algona, Iowa is located near the Minnesota border centered between the east and west borders of Iowa. Over 10,000 German POWs were held at the camp from April 1944 until February 1946. The POWs in Algona and its branch camps built and repaired fences, canned vegetables, processed hemp for rope production, and processed lumber. The POWs built a nativity scene for Christmas 1945 that is still displayed each year by the Algona POW museum, located in the area. After the camp closed in February 1946, the remaining POWs were sent to Ft. Crook, NE.

IA-1-1-1a

IA-1-1-5a

IA-1-1-10b

IA-1-1-25a

IA-1-1-50a

No.	Denomination	Paper	Ink	Serial	Block	Circ.	Unc.
IA-1-1-1a	1 cent	white	red	red	A, C, D	25	40
IA-1-1-1b	1 cent	white	blue	red	A	25	40
IA-1-1-1c	1 cent	white	green	red	A, D	25	40
IA-1-1-5a	5 cent	orange	red	red	A, C, D	35	55
IA-1-1-5b	5 cents	orange	blue	red	A	35	55
IA-1-1-5c	5 cents	orange	green	red			RNK
IA-1-1-10a	10 cents	white	red	red	A	55	85
IA-1-1-10b	10 cents	gray	blue	red	A	55	85
IA-1-1-10c	10 cents	tan	green	red	A, C	55	85
IA-2-1-10d	10 cents	tan	red	red	D	55	85
IA-1-1-25a	25 cents	pale orange	red	red	D	90	150
IA-1-1-25b	25 cents	pale orange	blue	red			RNK
IA-1-1-25c	25 cents	pale orange	green	red	D	90	150
IA-1-1-50a	50 cents	pink	red	red	D	125	200
IA-1-1-50b	50 cents	pink	blue	red			RNK
IA-1-1-50c	50 cents	pink	green	red			RNK

Clarinda

Clarinda, Iowa is located in the Southwestern corner of the state. The camp was opened on 293 acres on January 24, 1944. It was built for 3,000 POWs with 60 barracks and a 150 bed hospital. German POWs were first to arrive at the camp, followed by Italians and Japanese in 1945. POWs helped with farm labor, as well as repairing and sandbagging levees to prevent flooding. The camp closed on December 1, 1945, with the remaining POWs sent to California

IA-4-1-1a

IA-4-1-1b

No.	Denomination	Paper	Ink	Serial	Block	Circ.	Unc.
IA-4-1-1a	1 cent	green	blue	red		125	250
IA-4-1-1b	1 cent	tan	blue	red	B, D	100	200

KANSAS

There were 3 base camps located in Kansas—Concordia, Phillips, and Ft. Riley. These camps also served 10 branch camps in Kansas. Atlanta, NE and Ft. Crook, NE also served as base camps for several branch camps in Kansas. In addition, Leavenworth prison also held POWs for disciplinary reasons.

	Camp	Base Camp
1	*Camp Phillips	
2	Cawker City	Atlanta NE
3	Concordia	
4	Council Grove	Fort Riley
5	El Dorado	Fort Riley
6	Elkhart	Camp Phillips
7	Eskridge	Fort Riley
8	*Fort Riley	
9	Lawrence	Fort Riley
10	Leavenworth USDB	
11	Neodesha	Fort Riley
12	Ottawa	Fort Riley
13	Peabody	Fort Riley
14	Wadsworth	Fort Riley
15	Winter G.H.	Fort Crook NE

Concordia POW Base Camp

Concordia, Kansas is located 80 miles north of Salina near the Nebraska border. Camp Concordia was the largest of the camps in Kansas. It housed 4,000 German POWs that were captured in North Africa. The Camp was built in 90 days, opened in July 1943, and closed in November 1945. It had 304 buildings, including a 177 bed hospital and 815 guards. A museum recently opened on the site of the camp.

Phillips POW Base Camp

Camp Phillips was located near Salina, Kansas. This large military training camp, which housed 75,000 to 80,000 soldiers at a time, served as a POW camp for 3,000 Germans and Italians. The camp opened June 1943 and closed May 1946.

KS-1-1-1

No.	Denomination	Paper	Ink	Serial	Block	Circ.	Unc.
KS-1-1-1	1 cent	tan	blue	red		150	300

Ft. Riley POW Base Camp

Ft. Riley is located between Junction City and Manhattan, Kansas and was named after Major General Bennett C. Riley, the first military escort along the Santa Fe trail. It was home to the 2nd cavalry and the 9th armored division. Over 4,500 POWs from Germany, Italy, and Japan were held at Ft. Riley from 1943 until 1946. They worked on roads, building maintenance, the laundry, and local farms.

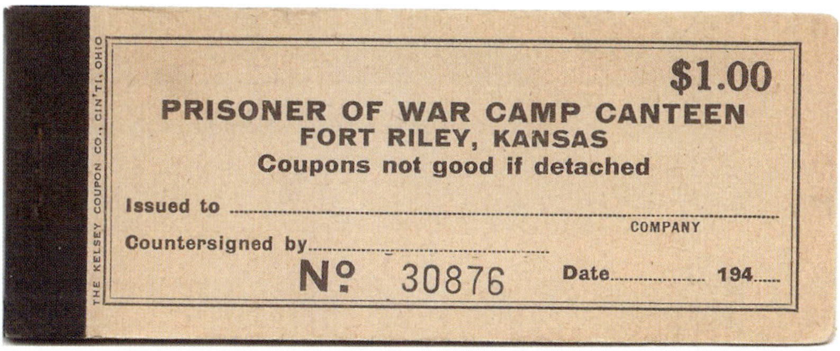

KS-8-1-1a

KS-8-1-1b

KS-8-1-5b

KS-8-2-1a

KS-8-2-5a

KS-8-3-10

No.	Denomination	Paper	Ink	Serial	Block	Circ.	Unc.
	Denomination on Right						
KS-8-1-1a	1 cent	peach	blue	red		25	40

KS-8-1-1b	1 cent	red	blue	black		25	40
KS-8-1-5a	5 cents	green	blue	red		35	55
KS-8-1-5b	5 cents	tan	blue	red		35	55
	Denomination on Left						
KS-8-2-1a	1 cent	peach	blue	red	A	25	40
KS-8-2-1b	1 cent	red	blue	red			RNK
KS-8-2-5a	5 cents	yellow	blue	red	A	35	55
KS-8-2-5b	5 cents	green					RNK
KS-8-3-10	10 cents	brown	black	red		150	300

KENTUCKY

Kentucky had three POW base camps—Breckenridge, Camp Campbell and Ft. Knox. Together they also supported 11 branch camps in the state.

	Camp	Base Camp
1	Bowman	Fort Knox
2	Breckenridge	
3	*Camp Campbell	
4	Danville	Fort Knox
5	Darnall G.H.	Fort Knox
6	Eminence	Fort Knox
7	*Fort Knox	
8	Frankfort	Fort Knox
9	Lexington	Fort Knox
10	Lexington	Fort Knox
11	Lexington Med. Dep.	Fort Knox
12	Lexington Sig. Dep.	Fort Knox
13	Maysville	Fort Knox
14	Owensboro	Breckenridge
15	Shelbyville	Fort Knox

Camp Breckenridge POW Base Camp

Camp Breckenridge is located about 40 miles southwest of Louisville. The camp started as an army post in 1942 and was named after John C. Breckinridge, a U.S. Vice President from 1856-1860. It was used from 1943-1946 as a POW camp for 3,000 German enlisted men. A POW named Daniel Mayer painted a 20 x 30 foot mural of a baroque castle, which is still present, in the only building that remains today.

Camp Campbell POW Base Camp

Camp Campbell, which is located near Hopkinsville, was authorized on September 23, 1942, and POWs began arriving in December 1942. It housed 1,804 German POWs until it closed in May 1946. The camp had a chapel, barbershop, school, a 250 seat theater, gymnasium, library, and hospital. The POWs had gardens and craft shops. Like other camps, local residents thought the POWs had better living standards than they did.

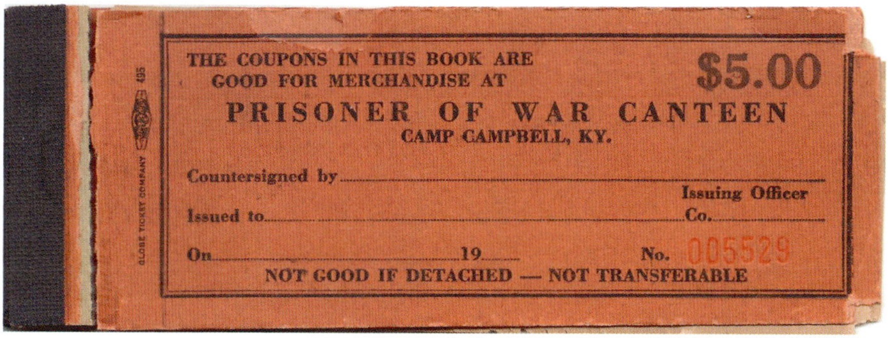

KY-3-1-5

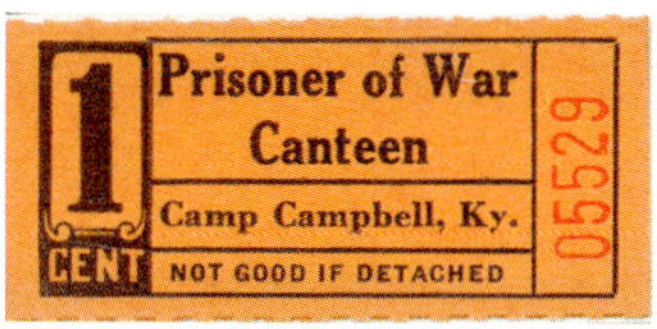

KY-3-2-1

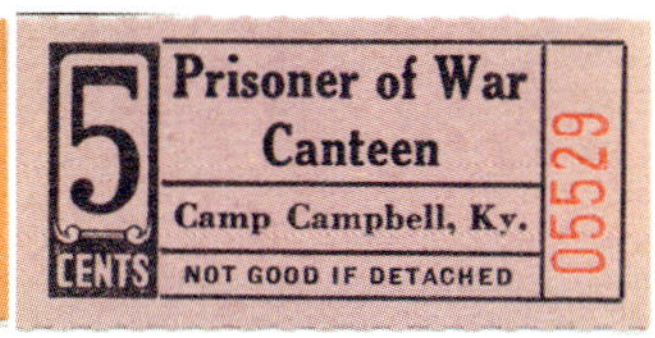

KY-3-2-5

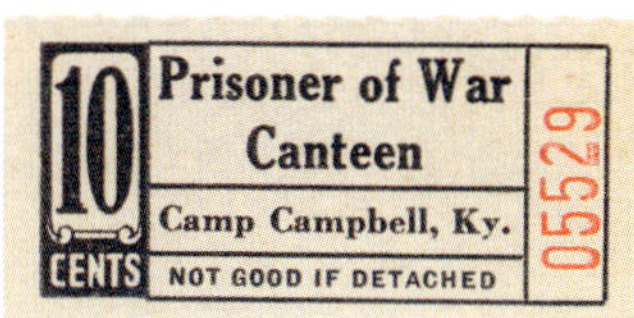

KY-3-2-10

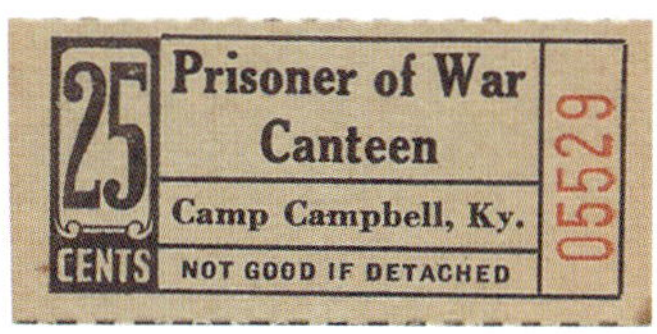

KY-3-2-25

No.	Denomination	Paper	Ink	Serial	Block	Circ.	Unc.
KY-3-1-5	5 cents Camp Campbell Exchange PW on back	mauve	blue	red	A	150	300
KY-3-2-1	1 cent	orange	black	red		125	250

KY-3-2-5	5 cents	tan	black	red		150	300
KY-3-2-10	10 cents	buff	black	red		150	300
KY-3-2-25	25 cents	buff	black	red		175	350

Ft. Knox POW Base Camp

Ft. Knox POW camp, located southwest of Louisville, was opened in February 1944 and held German POWs. The POWs at the camp worked in the tobacco fields and saved the loss of the crops due to a shortage of labor in 1943. There was an incidence of prisoner on prisoner violence that resulted in guards having to shoot seven prisoners, two dying from their wounds. A dozen graves at the camp contained POWs who were shot attempting to escape and others killed by Nazis to keep them in line. Many of the POWs enjoyed their stay at the camp and did not want to return to Germany after the war.

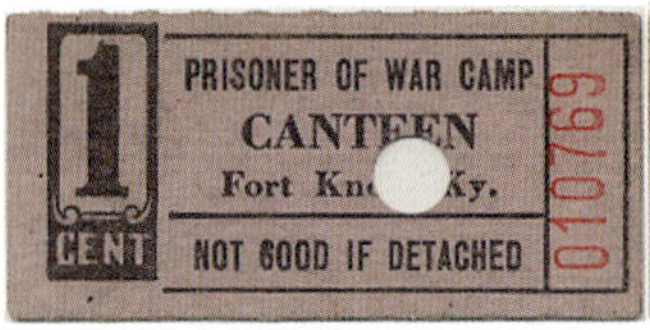

KY-7-1-1

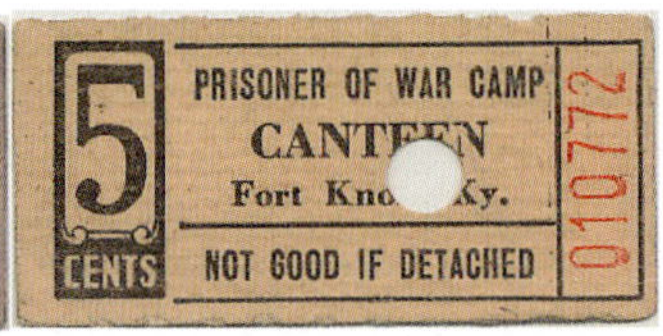

KY-7-1-5

KY-7-2-1a

KY-7-2-1b

KY-7-2-5

No.	Denomination	Paper	Ink	Serial	Block	Circ.	Unc.
KY-7-1-1	1 cent hole cancel	gray	black	red		125	250
KY-7-1-5	5 cent hole cancel	beige	black	red		125	250
KY-7-2-1a	1 cent	cream	green	red	B	125	250
KY-7-2-1b	1 cent	purple	blue	red	C	125	250
KY-7-2-5	5 cents	pale orange	blue	red	C	125	250
KY-7-2-10	10 cents	pale pink	blue	red	C	150	300
KY-7-2-25	25 cents	tan	blue	red	C	175	350

LOUISIANA

Louisiana was a significant state for the War Department's operation of prisoner of war camps in the United States. There were 44 camps in operation in the state, with base camps located at Camp Claiborne, Camp Livingston, Camp Polk, and Ruston.

	Camp	Base Camp
1	Arabi	Camp Livingston
2	Barksdale Field	Ruston
3	Bastrop	Ruston
4	*Camp Claiborne	
5	*Camp Livingston	
6	*Camp Polk	
7	*Camp Plauche	
8	Donaldsonville	Camp Livingston
9	Edgerly	Camp Livingston
10	Eunice	Camp Polk
11	Gueydan	Camp Polk
12	Hahnville	Camp Livingston
13	Hammond	Camp Livingston
14	Houma	Camp Livingston
15	Iowa	Camp Livingston
16	Jeanerette	Camp Polk
17	Jennings	Camp Polk
18	Jonesville	Camp Livingston
19	Kaplan	Camp Polk
20	Lake Charles AAF	Camp Polk
21	Lake Providence	Ruston
22	Lockport	Camp Livingston
23	Mansfield	Camp Polk
24	Mathews	Camp Livingston
25	Marksville	Camp Livingston
26	Melville	Camp Livingston
27	McCain GIN	Ruston

28	New Orleans AAB	Camp Plauche
29	New Orleans POE	
30	Point A La Hache	Camp Livingston
31	Port Allen	Camp Livingston
32	Port Sulphur	Camp Livingston
33	Rayne	Camp Livingston
34	Reserve	Camp Livingston
35	*Ruston	
36	St. Martinsville	Camp Polk
37	Simesport	Camp Livingston
38	Tallulah	Ruston
39	Terrebone	Camp Livingston
40	Thibodaux	Camp Livingston
41	Welsh	Camp Polk
42	West Monroe	Ruston
43	White Hall	Camp Livingston
44	Woodlawn Planta-	Camp Livingston
45	Youngsville	Camp Polk

Camp Claiborne

Camp Claiborne was located in central Louisiana in Rapides Parish. The camp was established in 1930 as Evangeline, with the name later changed to Claiborne after the first Governor of Louisiana, William C.C. Claiborne. The camp was used from 1939-1946 for basic training and artillery practice. Construction of the POW camp began in December 1944 and housed German POWs for only 5 months, from January-May 1945. Most of the POWs were used by companies in the area under contract with the Department of Defense for the harvesting of pulp wood. The main problem for the fair-skinned Germans was the abundance of poison ivy in the area.

LA-4-1-5

LA-4-1-10

LA-4-2-10

No.	Denomination	Paper	Ink	Serial	Block	Circ.	Unc.
	Camp Claiborne Exchange						
LA-4-1-5	5 cents	pale green	blue	red	B	125	250
LA-4-1-10	10 cents	orange	blue	red	C	125	250
	Camp Canteen						
LA-4-2-10	10 cents	tan	blue	red	C	175	350

Camp Livingston

Camp Livingston was a United States Army military camp that was named after Robert R. Livingston, a negotiator for the Louisiana Purchase. The camp, which was first known as Camp Tioga, was open from 1940 until 1945 and was located near Pineville in northern Louisiana. It was home to various infantry, combat engineer, and artillery units. The POW camp held 2,300 German, Italian, and Japanese prisoners from 1943 until 1945. One of the first POWs was a Japanese survivor of a three man midget submarine that attacked Pearl Harbor on December 7, 1941. The POWs worked primarily in logging and farm labor in the area.

LA-5-1-1a

LA-5-1-1b

LA-5-1-1c

LA-5-1-5

LA-5-1-10

No.	Denomination	Paper	Ink	Serial	Block	Circ.	Unc.
LA-5-1-1a	1 cent	pale pink	black	red	A	125	250
LA-5-1-1b	1 cent	pale pink	black	black	C	125	250
LA-5-1-1c	1 cent	white	black	red	A	125	250
LA-5-1-5	5 cents	yellow	black	black	C	150	275
LA-5-1-10	10 cents	gray	black	red	B	175	350

Camp Polk

Camp Polk was named after Reverand Leonidas Polk, a distinguished Confederate general from Louisiana. Construction began in 1940 for the camp, located in Vernon Parish, Louisiana. It is still in use today, known as Ft. Polk. It has been used as a basic training facility, as well as for infantry, combat support hospital, and armored cavalry divisions. Beginning in May of 1943, the camp housed over 3,000 German POWs. Their primary work was picking cotton, and cutting rice and lumber. They also helped to sandbag the flooded Red River area. The POW camp was closed in June 1946.

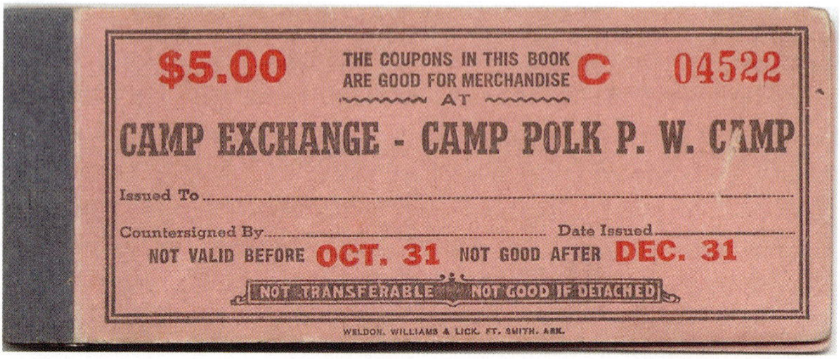

LA-6-1-5

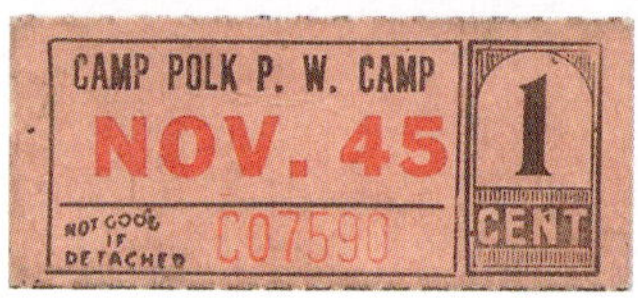

LA-6-2-1

LA-6-3-1

LA-6-3-5

No.	Denomination	Paper	Ink	Serial	Block	Circ.	Unc.
	PW Camp						
LA-6-1-5	5 cents	buff	black	red	D	150	300
	Nov 45 in Red						
LA-6-2-1	1 cent	pink	black	red	C	125	250
	Camp Polk Exchange						
LA-6-3-1	1 cent	pale brown	red	red	P	55	125
LA-6-3-5	5 cents	pink	blue	red	D	55	125

Camp Ruston

Camp Ruston, located in Ruston, Louisiana, was built in 1942 and was first used as a Women's Army Corps facility until June 1943. The camp was then operated as a POW camp until June 1946, housing 4,315 POWs. In addition to members of the German Afrika Korps, captured officers and crew of U-505 were held at the camp to prevent the Germans from discovering that secret naval codes had fallen to the Allies. French, Austrian, Italian, Czech, Polish, Yugoslav, Romanian, and Russian prisoners were held at the camp in addition to Germans. There were 34 POWs that managed to escape from Camp Ruston, with all but one recaptured. More information about the camp is available in the special collections department of Louisiana Tech University.

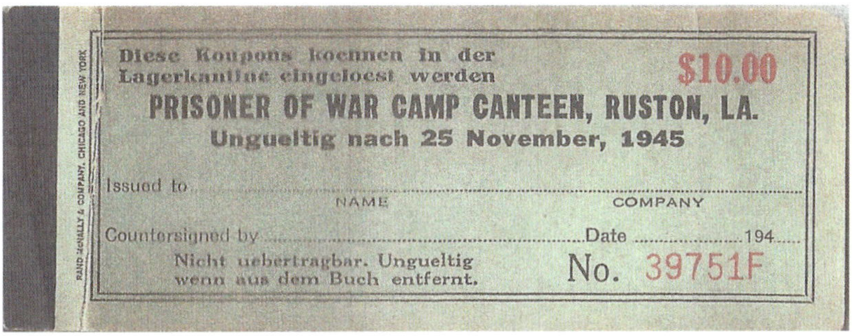

LA-35-1-1

LA-35-1-5

LA-35-1-10

LA-35-2-1

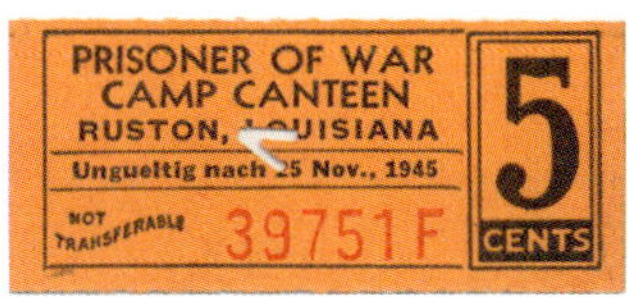

LA-35-2-5

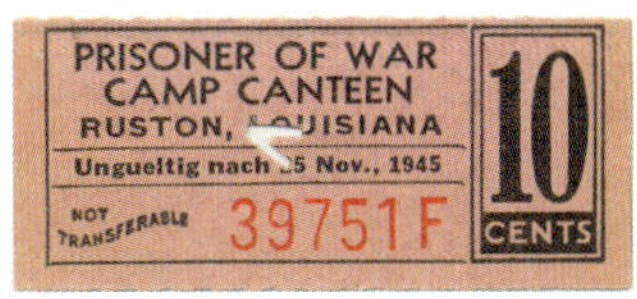

LA-35-2-10

LA-35-2-25

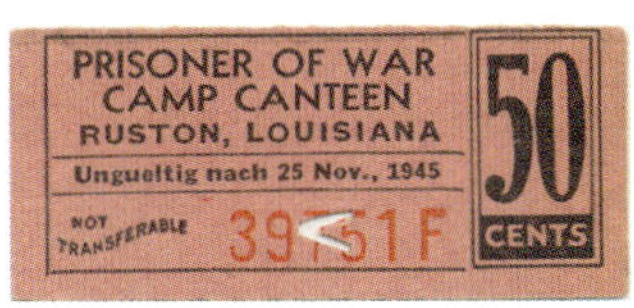

LA-35-2-50

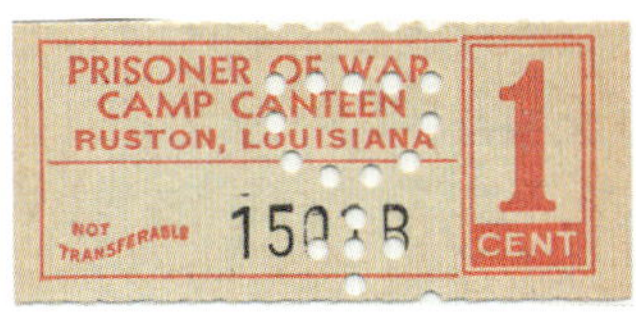

LA-35-3-1

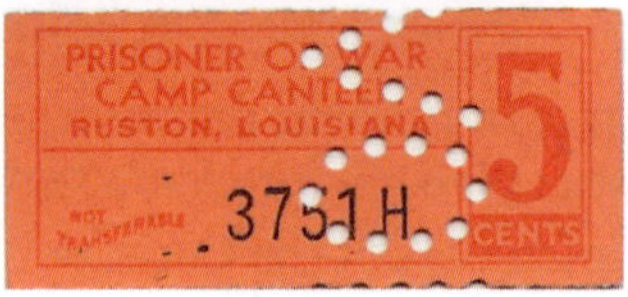

LA-35-3-5

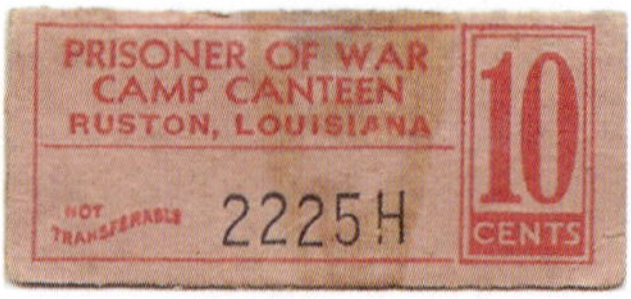

LA-35-3-10

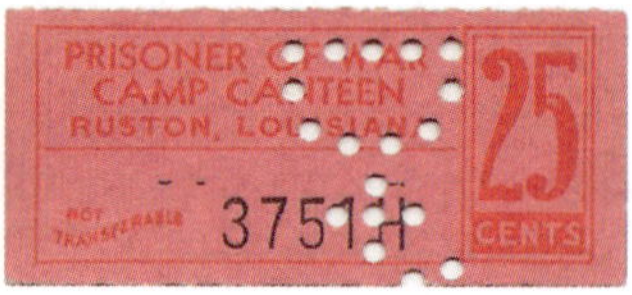

LA-35-3-25

LA-35-3-50

No.	Denomination	Paper	Ink	Serial	Block	Circ.	Unc.
LA-35-1-1	1 cent	buff	blue	red	A	25	40
LA-35-1-5	5 cents	pink	blue	red	A	35	55
LA-35-1-10	10 cents	buff	blue	red	A	55	85
	Ungueltig nach 25 Nov., 1945						
LA-35-2-1	1 cent	pale green	black	red	F	55	100
LA-35-2-5	5 cents	orange	black	red	F	55	100
LA-35-2-10	10 cents	pale rose	black	red	F	55	100
LA-35-2-25	25 cents	pink	black	red	F	100	200
LA-35-2-50	50 cents	pink	black	red	F	200	400
	Without expiration date						
LA-35-3-1	1 cent	pale green	red	black	B,H	25	40
LA-35-3-5a	5 cents	orange	red	black	B,H	35	55
LA-35-3-5b	5 cents	orange	black	black	S	55	85
LA-35-3-10	10 cents	pale rose	red	black	B,H	55	85
LA-35-3-25	25 cents	pink	red	black	G,H	90	150
LA-35-3-50	50 cents	pink	red	black	H	200	400

Camp Plauche (New Orleans)

Camp Plauche, located on the Western outskirts of New Orleans near the Huey P. Long bridge over the Mississippi, was formally known as Camp Harahan. It was renamed Camp Plauche after Major Jean Baptiste Plauche, who served under Andrew Jackson during the war of 1812. The camp was first used as a staging area for troops and then as an army training facility. In 1943 Camp Plauche became a POW camp, holding over 3,000 Italian POWs and members of the German Afrika Corps. The camp was closed in July 1946. A picture of a 5 cent chit has been seen.

MAINE

Maine had at least six POW camps, with Camp Holton being the only base camp and the others operating as sub-camps.

	Camp	Base Camp
1	Bangor	Houlton
2	*Houlton	
3	Presque Isle	Houlton
4	Princeton	Houlton
5	Seboemook	Houlton
6	Spencer Lake	Houlton

Camp Houlton

Camp Holton is located in eastern Maine on the US and Canadian border. The camp was the site of an Army airbase and, in 1944, became a POW camp that housed 3,700 German POWs. The POWs primarily picked potatoes and peas at farms in the area, chipped ice, and harvested lumber. The camp was closed in May 1946.

ME-2-1-1

ME-2-1-5

ME-2-2-5

ME-2-2-10

No.	Denomination	Paper	Ink	Serial	Block	Circ.	Unc.
ME-2-1-1	1 cent	pale yellow	blue	red	F	125	250
ME-2-1-5	5 cents	red	blue	red	E	125	250
	Number 2 upper left corner						
ME-2-2-5	5 cents	pink	blue	red	2 G	125	250
ME-2-2-10	10 cents	pink	blue	red	2 G	150	250

MARYLAND

During World War II, Maryland housed over 10,000 POWs in 20 base and subcamps. The first Italian POWs came to Maryland in September 1943 and were sent to Ft. George G. Meade. By 1945, 4,000 POWs were working for the Army or Navy, while 6,000 worked for civilian contractors. Many of the POWs worked in camp bakeries, canteens, hospitals, and laundries, with others engaged in farm labor and road construction. It is estimated that POW labor saved Maryland over 5 million dollars in agricultural labor from June-December 1945.

	Camp	Base Camp
1	Andrews	Fort George G. Meade
2	Berlin	Custis
3	Cambridge	Custis
4	Church Hill	Custis
5	Easton	Custis
6	Edgewood ARS	
7	Flintstone	Fort George G. Meade
8	*Fort George G. Meade	
9	Frederick	Fort George G. Meade
10	Gaithersburg	Fort George G. Meade
11	Holabird Sig. Dep.	
12	Hurlock	Custis
13	Logan Field	Holabird
14	Nanjemoy	Fort George G. Meade
15	Pikesville	Fort George G. Meade
16	Ritchie	
17	Somerset	Custis
18	Fort Washington	Fort George G. Meade
19	Westminster	Fort George G. Meade

Camp Holabird

Camp Holabird was established in 1917 and is located 2-3 miles southeast of Baltimore on the Patapsco River. The camp housed approximately 750 German POWs with the adjacent Logan Field holding 450 Italian POWs. The camp operated from October 1944 until May 1946. The Italian POWs were allowed to work in areas off the base, while the Germans were not.

Ft. George G. Meade

Ft. Meade was established by the US Army in 1917 as a training facility for troops in World War I. The fort is located in Anne Arundel County and was named after General George Meade, in honor of the victory at the battle of Gettysburg in 1863. During World War II, the 76th and 29th Virginia/Maryland divisions trained there. In 1943 Italian POWs began arriving at the fort until May 1944, when it became a German POW camp. The camp held mostly Wehrmacht (army) personnel with a few Luftwaffe prisoners. There was, however, a highly decorated submarine commander named Werner Henke, who was shot and killed trying to escape.

MD-8-1-1a

MD-8-1-1b

MD-8-1-1c

MD-8-1-1d

MD-8-1-2a

MD-8-1-2b

MD-8-1-2c

MD-8-1-5a

MD-8-1-5b

MD-8-1-5c

MD-8-1-5d

MD-8-1-10a

MD-8-1-10b

MD-8-1-10c

MD-8-1-10d

MD-8-1-25a

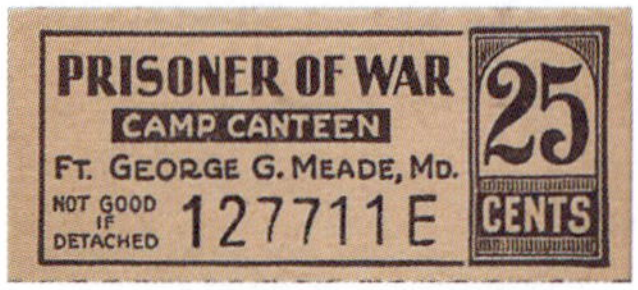

MD-8-1-25b

MD-8-1-25c

No.	Denomination	Paper	Ink	Serial	Block	Circ.	Unc.
MD-8-1-1a	1 cent	orange	black	purple	B	25	40
MD-8-1-1b	1 cent	cream	blue	red	C	25	40
MD-8-1-1c	1 cent	pink	black	black	A	25	40
MD-8-1-1d	1 cent	rose	black	black	D	25	40
MD-8-1-2a	2 cents	tan	black	purple	B	50	100
MD-8-1-2b	2 cents	pale green	black	black	E	50	100
MD-8-1-2c	2 cents	buff	black	black	E	50	100
MD-8-1-5a	5 cents	mauve	black	purple	A,B	35	55
MD-8-1-5b	5 cents	orange	black	red	C	35	55
MD-8-1-5c	5 cents	pink	black	black	D,E	35	55
MD-8-1-5d	5 cents	buff	black	black	E	35	55
MD-8-1-10a	10 cents	purple	blue	red	C	55	85
MD-8-1-10b	10 cents	pale yellow	black	purple	A	55	85
MD-8-1-10c	10 cents	pale green	black	purple	B	55	85
MD-8-1-10d	10 cents	pink	black	black	D	55	85
MD-8-1-25a	25 cents	pink	black	black	D	90	150
MD-8-1-25b	25 cents	buff	black	black	E	90	150
MD-8-1-25c	25 cents	yellow	black	purple	B	90	150

Edgewood Arsenal

Edgewood Arsenal, established in 1917, was a camp used for ordnance testing and chemical and biological warfare. Today it is called the Aberdeen Proving Grounds. Little is known about the camp except that it operated from October 1944 until May 1946 and held 700 German POWs.

Camp Ritchie

Camp Richie was built by the Maryland National Guard in 1926 and named after Albert C. Richie, the Maryland governor at the time. In 1942, the camp was taken over by the War Department to be used for training because it had weather conditions similar to Germany. In November 1944, Japanese Nisei women were brought into the camp to be used in intelligence, working with captured Japanese documents. There was a POW camp established in 1944 where approximately 200 German POWs were held. Another interesting fact about Ft. Richie is that some German Jewish refugees trained there and were used to interrogate POWs both in the camp and in Europe.

Camp Somerset

Camp Somerset is located in Westover, Maryland and began in 1935 as a CCC camp. The camp was taken over in January 1942 for costal defense shore protection. Due to a shortage of farm labor, Somerset County requested German POWs, and Camp Somerset became a POW camp in August 1944. The POWs worked in canning houses, sawmills, and meat packing plants in the area. The camp closed in June 1946, having held 1,450 POWs. Today the camp is a migrant workers facility.

MASSACHUSETTS

There were at least eight POW camps in Massachusetts that held both German and Italian POWs. They operated from March 1944 until June 1946. The camps held over 11,000 POWs during their existence. Information as to which were base and subcamps is unclear. However, it is thought that Boston Port of Entry, Ft. Devens, Ft, Myles Standish, Ft. Strong, and Westover Field were considered base camps. At present only Ft. Devens and Ft. Edwards chits are known to exist.

	Camp or Base Camp
1	Boston POE
2	*Camp Edwards
3	*Fort Devens
4	Fort Strong
5	Lowel G.H.
6	McKay
7	Myles Standish
8	Westover Field

Ft. Devens

Ft. Devens is located in Middlesex County and was established in 1917. It was named after Charles Devens, a general in the Union Army during the Civil War. During World War II three army divisions were trained there, and a POW camp held over 3,000 German POWs.

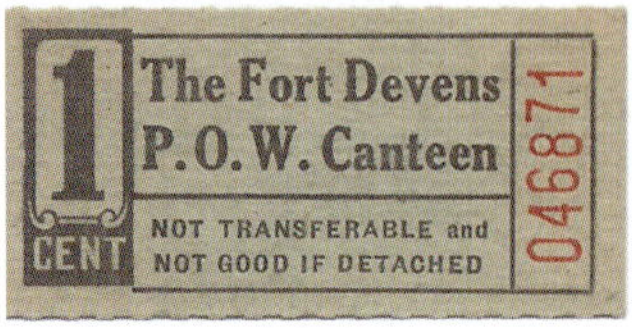

MA-3-1-1a

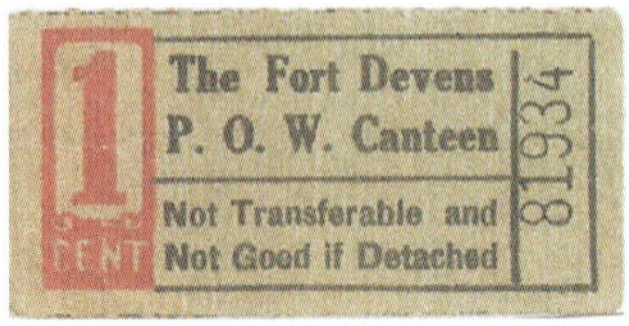

MA-3-1-1b

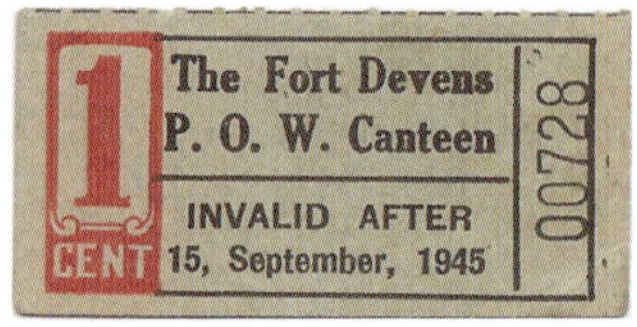

MA-3-2-1

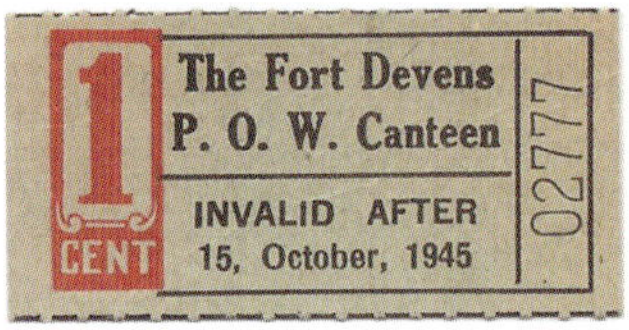

MA-3-3-1

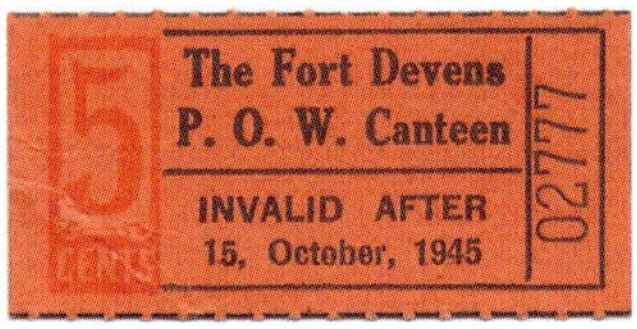

MA-3-3-5

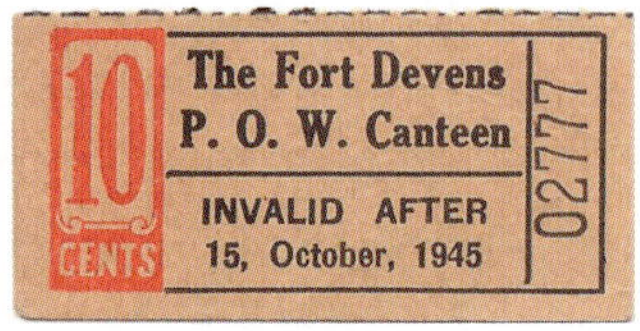

MA-3-3-10

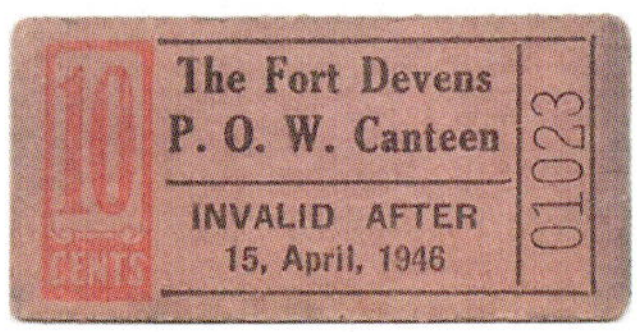

MA-3-4-10

No.	Denomination	Paper	Ink	Serial	Block	Circ.	Unc.
MA-3-1-1a	1 cent	gray	black	red		125	250
MA-3-1-1b	1 cent	gray	red & black	black		125	250
	Invalid after 15 September 1945						
MA-3-2-1	1 cent	gray	red & black	black		125	250
	Invalid after 15 October 1945						
MA-3-3-1	1 cent	gray	red & black	black		125	250
MA-3-3-5	5 cents	orange	red & black	black		150	300
MA-3-3-10	10 cents	pale yellow	red & black	black		175	350
	Invalid after 15 April 1946						
MA-3-4-10	10 cents	pink	red & black	black		175	350

Ft. Edwards POW Camp

MA-2-1-2

No.	Denomination	Paper	Ink	Serial	Block	Circ.	Unc.
MA-2-1-2	2 cents	buff	blue	red	C	150	300

MICHIGAN

There were 6,000 German POWs held in Michigan, with 1,000 in five camps in the Upper Peninsula. The base camp for the POW camps in the Upper Peninsula was actually not in Michigan but was Ft. Sheridan, Illinois. Many lumber and chemical companies benefited from the labor of the POWs. The camps in the rest of Michigan were operated out of one base camp at Ft. Custer.

	Camp	Base Camp
1	Allegan	Fort Custer
2	Alma	Fort Custer
3	AuTrain	Fort Sheridan IL
4	Bay City	Fort Custer
5	Benton Harbor	Fort Custer
6	Blissfield	Fort Custer
7	Cairo	Fort Custer
8	Coloma	Fort Custer
9	Croswell	Fort Custer
10	Dundee	Fort Custer
11	Evelyn	Fort Sheridan IL
12	*Fort Custer	
13	Freeland	Fort Custer
14	Fremont	Fort Custer
15	Grosse Isle	Fort Custer
16	Grant	Fort Custer
17	Hart	Fort Custer

18	Hartford	Fort Custer
19	Lake Odessa	Fort Custer
20	Pori (Mass)	Fort Sheridan IL
21	*Owasso	Fort Custer
22	Raco	Fort Sheridan IL
23	Mt. Pleasant	Fort Custer
24	Romulus AAF	Fort Custer
25	Shelby	Fort Custer
26	Sidnaw	Fort Sheridan IL
27	Sodus	Fort Custer
28	Waterloo	Fort Custer
29	Wayne	

Ft. Custer

Ft. Custer is located four miles west of Battle Creek and is used today as a Michigan Army National Guard training facility. The facility began as Camp Custer in 1917 and was named after General George Armstrong Custer, a Michigan native. In 1940, Camp Custer became a permanent training facility and became Ft. Custer. More than 300,000 troops trained there including the Red Diamond Division (5th Infantry). In 1943, a POW camp was created at Ft. Custer and contained 5,000 German POWs. They performed work on farms and vineyards in the area. On October 31, 1945, sixteen German POWs were killed when a truck returning from a work detail collided with a train. The POW camp was closed in May 1946.

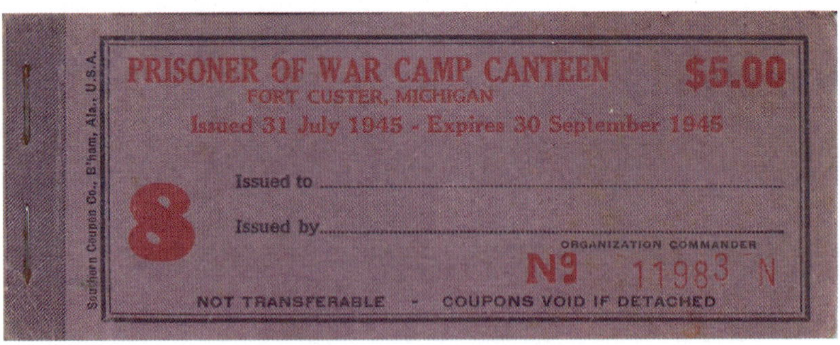

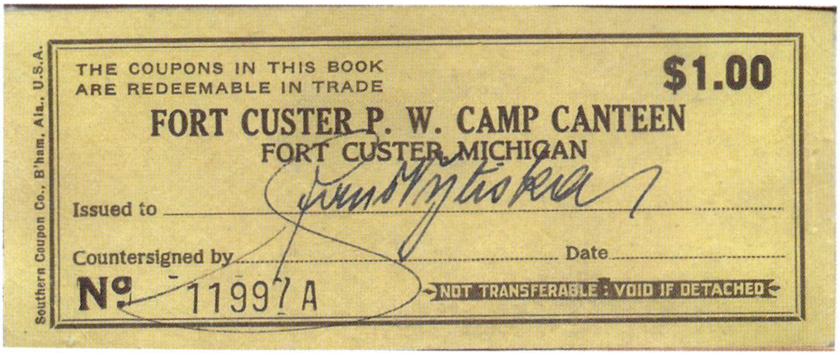

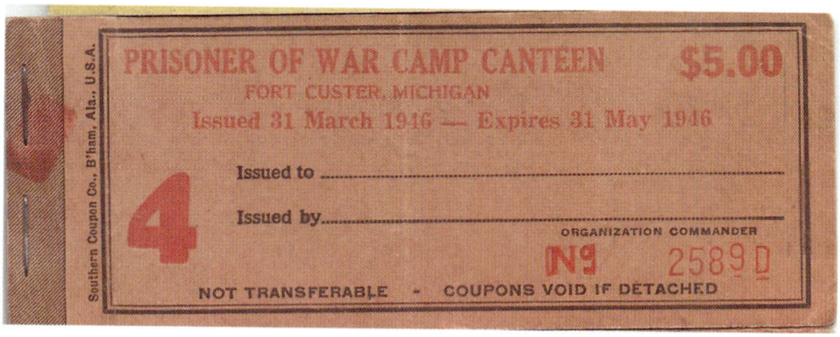

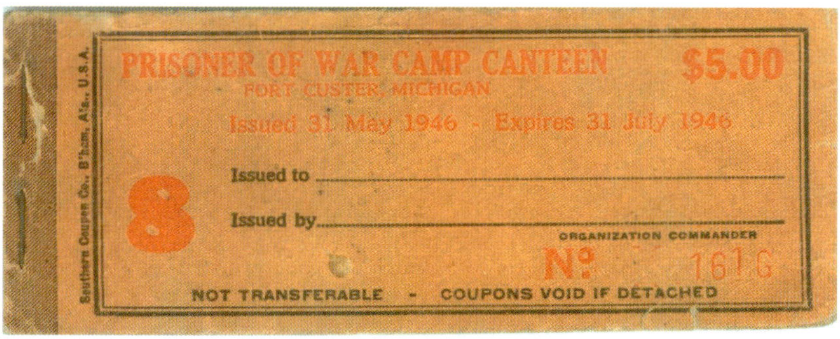

MI-12-1-1

MI-12-2-1

MI-12-2-2

MI-12-2-5

MI-12-3-1a

MI-12-3-1b

MI-12-3-1c

MI-12-3-2

MI-12-3-5a

MI-12-3-5b

MI-12-3-5c

No.	Denomination	Paper	Ink	Serial	Block	Circ.	Unc.
	Not good after 31 May 1945						
MI-12-1-1	1 cent	pale green	red	red	E	100	200
	Not good after 30 November						
MI-12-2-1	1 cent	green	blue	red	K	100	200
MI-12-2-2	2 cents	red	blue	red	H	100	200
MI-12-2-5	5 cents	buff	blue	red	H	100	200
	No date						
MI-12-3-1a	1 cent	mauve	blue	red	4 D	100	200
MI-12-3-1b	1 cent	mauve	blue	red	8 G	100	200
MI-12-3-1c	1 cent	blue	blue	red	8 N	100	200
MI-12-3-2	2 cents	pale green	blue	red	8 G	100	200
MI-12-3-5a	5 cents	tan	blue	red	8 G	125	250
MI-12-3-5b	5 cents	white	blue	red	8 N	125	250
MI-12-3-5c	5 cents	white	blue	red	7 M	125	250

Owosso POW Camp

Only one chit is known from this camp.

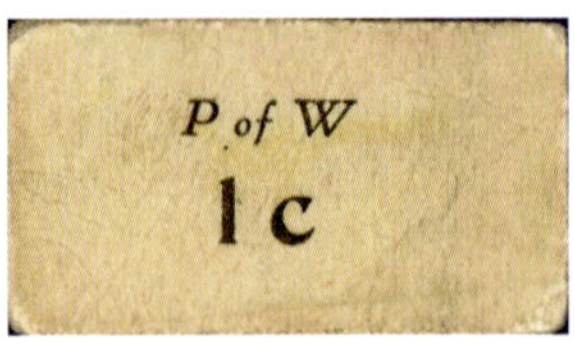

MI-21-1-1

No.	Denomination	Paper	Ink	Serial	Block	Circ.	Unc.
MI-21-1-1	1 cent	white	black			200	400

MINNESOTA

Minnesota had twenty-one branch camps holding over 6,000 German POWs. There were three base camps that directed the sub-camps, but none were in Minnesota. The three base camps were Algona, Iowa; Concordia, Kansas; and Ft. Crook, Nebraska. In northern Minnesota POWs harvested pulp wood, while POWs in central and southern Minnesota harvested crops and worked in the canning industry. There were very few problems with the Minnesota POWs and almost no escape attempts. The only problem occurred when POWs, who were captured later in the war, mainly young boys and older men, clashed with veteran POWs from the Afrika Korps.

	Camp	Base Camp
1	Ada	Algona IA
2	Bena	Algona IA
3	Bird Island	Algona IA
4	Crookston	Algona IA
5	Cut Foot Souix	Concordia KS
6	Deer River	Algona IA
7	Fairmont	Algona IA
8	Faribault	Algona IA
9	Grand Rapids	Algona IA
10	Hollandale	Algona IA
11	Montgomery	Algona IA
12	Moorhead	Algona IA
13	New Ulm	Algona IA
14	Olivia	Algona IA
15	Ortonville	Algona IA
16	Owatonna	Algona IA
17	Remer	Concordia KS
18	St. Charles	Algona IA
19	Warren	Fort Crook NE
20	Wells	Algona IA

MISSISSIPPI

German and Italian POWs began arriving in Mississippi soon after the campaign in North Africa was completed. Most of the POWs ranged in age from 18 to 20 when they were captured. There were four base POW camps, with over seventeen branch camps. Over 20,000 POWs were held in Mississippi with ¾ of them German and ¼ Italian. The POWs in camps in the delta chopped weeds and picked cotton, while the POWs in Southern Mississippi worked in the pine forests planting seedlings and harvesting lumber.

	Camp	Base Camp
1	Belzoni	Camp McCain
2	Brookhaven	Clinton
3	*Camp McCain	
4	*Camp Shelby	
5	Clarksdale	Camp McCain
6	*Clinton	
7	*Como	
8	Drew	Camp McCain
9	Elkas	Camp McCain
10	Greenville	Clinton
11	Greenwood	Camp McCain
12	Green Wood AAF	Camp McCain
13	Gulfport NTC	Camp Shelby
14	Indianola	Camp McCain
15	Lake Washington	Camp Shelby
16	Leland	Camp McCain
17	Merigold	Camp McCain
18	Picayune	Camp Shelby
19	Richton	Camp Shelby
20	Rosedale	Camp McCain
21	Saucier	Camp Shelby
22	Van Dorn	

Camp Clinton

Camp Clinton, located in Clinton, Mississippi, was the first POW camp to receive prisoners in August 1943. Many German generals and admirals were held at Camp Clinton, including General Von Arnim, Rommel's replacement in North Africa. The general was held at a private villa, had access to a car and driver, and was known to travel to Jackson for entertainment. One interesting project the POWs participated in was the Mississippi River Basin Model. An area of one square mile was cleared in the Jackson, Mississippi area by the POWs, under the supervision of the Corps of Engineers. An elaborate escape attempt also occurred at Camp Clinton. A 100 foot tunnel was dug and included bulbs to light the tunnel. However, the escaping POWs were captured only 10 feet from the fence.

MS-6-1-1

No.	Denomination	Paper	Ink	Serial	Block	Circ.	Unc.
MS-6-1-1	1 cent	rose	black	red	B	150	300

Camp Como

Camp Como was constructed as a POW camp by the Corps of Engineers in 1942 in Como, Mississippi. POWs started arriving there in September 1943. There were 3,800 Italian POWs first housed at the camp, but they were removed due to racial and ethnic problems with the German POWs. One of the oddest events involved a German POW pilot and the wife of a Mississipppi Delta planter. They were captured in Nashville while they were on their way to the East coast to steal an airplane and fly it to Greenland.

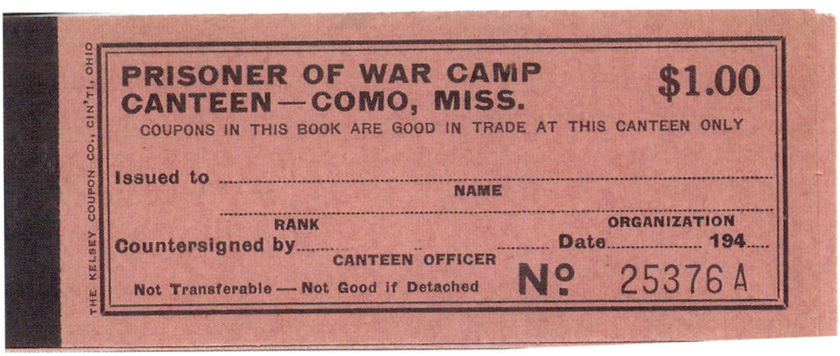

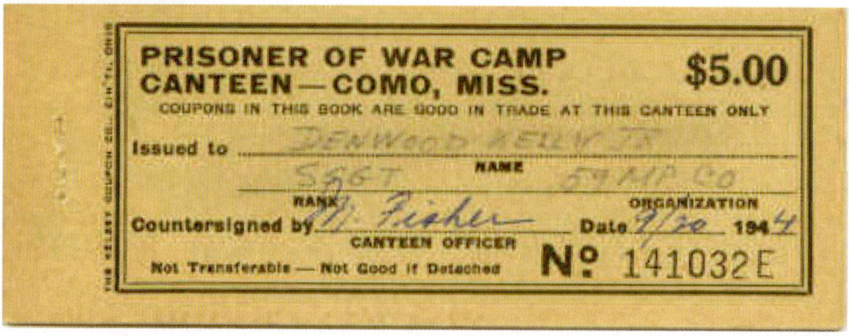

MS-7-1-1

MS-7-1-5

MS-7-1-10

MS-7-1-25

No.	Denomination	Paper	Ink	Serial	Block	Circ.	Unc.
MS-7-1-1	1 cent	pink	blue	red	A,E	25	40
MS-7-1-5	5 cents	pale orange	blue	red	A,E	35	55
MS-7-1-10	10 cents	pale tan	blue	red	A,E	55	85
MS-7-1-25	25 cents	brown	blue	red	A,E	90	150

Camp McCain

Camp McCain was built in 1942 near Elliot, Mississippi by the US Army to be used as a training facility and bombing range. The camp was named for a local Major General, Henry P. McCain. In 1944, the camp became a POW camp for captured German soldiers, holding 7,700. POWs were used in farming, mainly picking cotton. The POW camp was closed October 15, 1946. Today it is a National Guard training facility.

MS-3-1-1

MS-3-1-5

MS-3-2-5

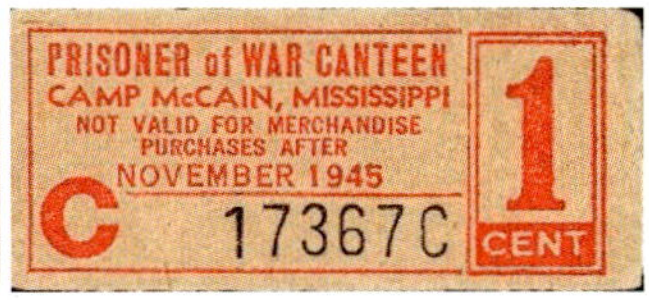

MS-3-3-1

MS-3-3-5

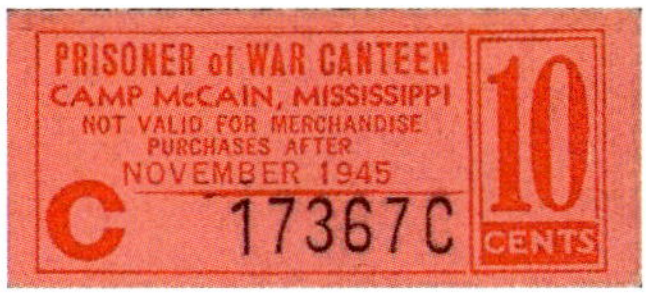

MS-3-3-10

MS-3-3-25

MS-3-3-1.00

MS-3-4-1

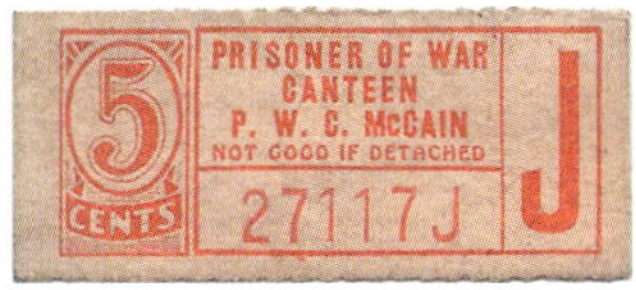

MS-3-4-5

No.	Denomination	Paper	Ink	Serial	Block	Circ.	Unc.
	Prisoner of 1st Line						
MS-3-1-1	1 cent	beige	black	red		150	300
MS-3-1-5	5 cents	orange	black	red		150	300
	May 30, 1945						
MS-3-2-5	5 cents	pale orange	red & black	red	B	150	300
	November 1945						
MS-3-3-1	1 cent	buff	red	black	C C	150	300
MS-3-3-5	5 cents	orange	red	black	C C	150	300
MS-3-3-10	10 cents	pink	red	black	C C	150	300
MS-3-3-25	25 cents	pale yellow	red	black	C C	175	350
MS-3-3-100	1 Dollar	gray	red	black	C C	400	800
	PWC Canteen & J						
MS-3-4-1	1 cent	white	red	red	J	150	300
MS-3-4–5	5 cents	pale tan	red	red	J	150	300

Camp Shelby

Camp Shelby is located in Hattiesburg, Mississippi and is currently the largest state owned and operated field training facility in the United States. The camp was established in 1917 and named after Revolutionary War hero and first Governor of Kentucky, Isaac Shelby. During World War II it was known for training the highly-decorated Japanese-American 442nd Regimental Combat Team. A POW camp was opened in August 1943 and held 5,300 German POWs until it closed in May 1946. In addition to forestry work, POWs helped to complete Lake Shelby, a small lake near the camp.

MS-4-1-1

MS-4-1-2

MS-4-1-5

MS-4-1-10

MS-4-1-25

No.	Denomination	Paper	Ink	Serial	Block	Circ.	Unc.
MS-4-1-1	1 cent	white	green	red	D	125	250
MS-4-1-2	2 cents	orange	green	red	D	175	350
MS-4-1-5	5 cents	rose	green	red	D	150	300
MS-4-1-10	10 cents	pale green	green	red	D	175	350
MS-4-1-25	25 cents	tan	green	red	D	200	400

MISSOURI

Missouri began receiving POWs in February 1943, starting with Camp Clark. Over 15,000 Italian and German POWs were held in four base camps and twenty subcamps, including six boat camps located on the Missouri and Mississippi rivers. The four base camps were Camp Clark, Camp Crowder, Ft. Leonard Wood, and Weingarten. The POWs generally worked in agriculture, harvesting potatoes, and shucking oats and wheat. Ft. Crowder and Ft. Leonard Wood were the last camps to close in May 1946.

	Camp	Base Camp
1	*Camp Clark	
2	*Camp Crowder	
3	Charleston	Weingarten, MO
4	Chesterfield I	Ft. Leonard Wood, MO

5	Chesterfield II	Ft. Leonard Wood, MO
6	Chesterfield Boat	Ft. Leonard Wood, MO
7	*Fort Leonard Wood	
8	Gasconade	Weingarten, MO
9	Glasgow	Ft. Leonard Wood, MO
10	Grand Pass	Ft. Riley, KS
11	Hannibal	Clarinda, IA
12	Independence	Ft. Leonard Wood, MO
13	Jefferson Baracks	Ft. Leonard Wood, MO
14	Kansas City	Camp Clark, MO
15	Kennettt	Weingarten, MO
16	Lexington	Ft. Riley, KS
17	Liberty	Ft. Riley, KS
18	Louisiana	Ft. Leonard Wood, MO
19	Maulden	Ft. Leonard Wood, MO
20	Marshal	Ft. Leonard Wood, MO
21	Marston	Ft. Leonard Wood, MO
22	New Madrid	Clarinda, IA
23	Orrick	Ft. Leonard Wood, MO
24	O'Reilly General Hospital	Weingarten, MO
25	Parksville	Weingarten, MO
26	Rosati	Ft. Leonard Wood, MO
27	St. Louis	Ft. Leonard Wood, MO
28	St. Louis ORD	Weingarten, MO
29	Sedalia AAF	Ft. Leonard Wood, MO
30	Sikeston	Weingarten, MO
31	Springfield	Camp Clark, MO
32	Washington	Ft. Leonard Wood, MO
33	*Weingarten	

Camp Clark

Camp Clark was established in 1908 in Nevada, Missouri. The camp was named for General Harvey C. Clark, commander of the Missouri National Guard. The camp was initially used for an assembly point for troops sent to Central America, the Mexican border wars, and World War I. In 1942, the camp was reopened as a POW camp for 5,000 German and Italian prisoners. The POW camp consisted of three compounds, each designed to hold approximately 1,000 prisoners, and also a 200-bed hospital. Two of the three compounds contained 48 structures that were subdivided into four companies. Each company consisted of eight barracks, a mess hall, a store/recreation hall, and a lavatory. One of the three compounds served as an Internment Camp and is possibly the last intact POW camp left in the country. Fanatical Nazis and trouble-makers from other camps were sent to Camp Clark. Presently, the camp is used for weekend and annual training and equipment storage by the Missouri National Guard.

MO-1-1-5

MO-1-2-1

MO-1-2-5

MO-1-2-10

MO-1-2-25

No.	Denomination	Paper	Ink	Serial	Block	Circ.	Unc.
	Good in trade						
MO-1-1-5	5 cents	orange	red	black	S	150	300
	Camp Canteen						
MO-1-2-1	1 cent	rose	blue	red	A	25	40
MO-1-2-5	5 cents	pale orange	blue	red	A	35	55
MO-1-2-10	10 cents	pale yelow	blue	red	A	55	85
MO-1-2-25	25 cents	brown	blue	red	A	90	150

Camp Crowder

Camp Crowder was named for Major General Enoch H. Crowder, who was provost marshal during World War I and creator of the Selective Service Act of 1917. The camp was opened in 1941 near Neosho, Missouri as a signal corps training camp.

Later, an officers candidate training and medical corps training facility were created. A POW camp was opened at Camp Crowder in February 1944 to house 2,000 German POWs. The camp closed in May 1946.

MO-2-1-1

MO-2-2-1

No.	Denomination	Paper	Ink	Serial	Block	Circ.	Unc.
	Camp Crowder Exchange PW						
MO-2-1-1	1 cent	buff	black	red	B	175	350
	Camp Canteen						
MO-2-2-1	1 cent	pale gray	red	black	B	175	350

Ft. Leonard Wood

Construction for Ft. Leonard Wood, one of the nation's largest training facilities, began in December 1940. Ft. Leonard Wood is located in south central Missouri between the towns of St. Roberts and Waynesville. The fort was named after Major General Leonard Wood, commander of the Rough Riders during the Spanish American War. During World War II, over 300,000 soldiers were trained at Ft. Leonard Wood. In June 1943, a POW camp was opened to house Italian and German POWs. POWs worked on farms detasseling corn and harvesting potatoes, fruit orchards, and nurseries. Few escapes took place, as many of the POWs thought conditions were much better than at home. The camp closed in June 1946. Today Ft. Leonard Wood is still a large army training facility and has three excellent museums representing areas of training (engineers, chemical units, and MPs).

MO-7-1-1

MO-7-1-5

MO-7-2-1

MO-7-2-25

No.	Denomination	Paper	Ink	Serial	Block	Circ.	Unc.
	Exchange						
MO-7-1-1	1 cent	pale orange	blue	red	F	150	300
MO-7-1-5	5 cents	pale orange	blue	red	F	150	300
	Canteen						
MO-7-2-1	1 cent	buff	black	black		175	350
MO-7-2-25	25 cents	rose	black	black		175	350

Weingarten

Weingarten is a small unincorporated area between Ste. Genevieve and Farmington, Missouri on State Highway 32. The camp consisted of 950 acres and some 380 buildings of various types and had a capacity of 5,800. The servicemen who would run the camp began arriving in November 1942, with the first prisoners arriving May 1943. The camp housed about 5,000 Italian prisoners. The Italian POWs were well-liked and very cooperative. The POWs asked permission to set up their own mess hall and soon had the guards eating there because of the more appetizing dishes. Since St. Louis had a large Italian population, family visits between civilians and POWs took place. Only two escapes occurred, in which the escapees were both quickly caught. The escape was due to boredom at the camp. The camp was closed in October 1945, and, today, only a chimney remains.

MO-33-1-1

MO-33-1-25

No.	Denomination	Paper	Ink	Serial	Block	Circ.	Unc.
MO-33-1-1	1 cent	buff	blue	red	C	150	300
MO-33-1-25	25 cents	pale orange	blue	red	C	175	350

MONTANA

Montana had 15 branch POW camps but no base camp located in the state. The base camp for Montana's camps was Rupert, Idaho. All of the Montana camps were open from September 1944 until November 1944 and again from July 1945 until they closed in September 1945. POWs in Montana primarily harvested beets and worked in the sugar plants. There were both Italian and German POWs held in Montana, probably numbering between five and six thousand.

	Camp	Base Camp
1	Ballentine	Rupert, ID
2	Billings	Rupert, ID
3	Bridger	Rupert, ID
4	Chinook	Rupert, ID
5	Forsythe	Rupert, ID
6	Fort Shaw	Rupert, ID
7	Glasgow	Rupert, ID
8	Hardin I	Rupert, ID
9	Hardin II	Rupert, ID
10	Harlem	Rupert, ID
11	Holly	Rupert, ID
12	Laurel	Rupert, ID
13	Malta	Rupert, ID
14	Miles City	Rupert, ID
15	Sidney	Rupert, ID

NEBRASKA

Nebraska held about 12,000 German and Italian POWs beginning in mid-1943 and ending in mid-1946. There were 5 base camps at Atlanta, Ft. Crook, Ft. Robinson, Indianola, and Scottsbluff and 24 sub-camps. POWs were very helpful in picking Nebraska's record corn crop of 1944.

	Camp	Base Camp
1	Alma	Atlanta, NE
2	Arapahoe	Indianola, NE
3	*Atlanta	
4	Bayard	Scottsbluff, NE

5	Bertrand	Atlanta, NE
6	Bridgeport	Scottsbluff, NE
7	Cornhusker	Atlanta, NE
8	Elwood	Atlanta, NE
9	Fishers Farm	Indianola, NE
10	*Fort Crook	
11	*Fort Robinson	
12	Franklin	Atlanta, NE
13	Grand Island	Atlanta, NE
14	Hastings	Atlanta, NE
15	Hayes Center	Indianola, NE
16	Hebron	
17	*Indianola	
18	Kearney	Atlanta, NE
19	Knobel	Ft. Robinson, NE
20	Lexington	Atlanta, NE
21	Lyman	Scottsbluff, NE
22	Mitchell	Scottsbluff, NE
23	Ogalalla	Indianola, NE
24	Palasade	Indianola, NE
25	*Scottsbluff	
26	Sidney	Scottsbluff, NE
27	Turrell	Ft. Robinson, NE
28	Weeping Water	Atlanta, NE
29	West Point	Clarinda, IA

Atlanta

Construction began on the camp just outside of the town of Atlanta, Nebraska in September 1943. In December 1943, 250 German POWs arrived followed by 844 more in February 1944. The camp consisted of a chapel, theater, post exchange, laundry, hospital, bakery, and repair shops. More POWs were also transferred to the Atlanta camp from Camp Concordia, Kansas. Besides helping local farmers with planting and harvesting, the camp had a large bakery which produced 522 loaves of white and rye bread daily. At its peak, the camp held over 3,000 POWs and was closed in March 1946.

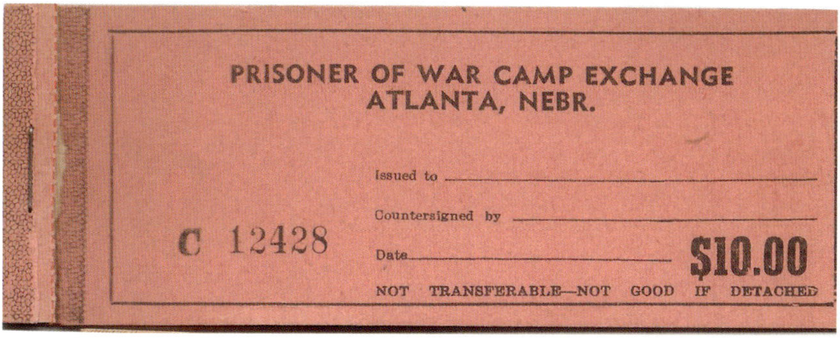

NE-3-1-1

NE-3-1-5

NE-3-1-10

NE-3-1-25

NE-3-1-50

No.	Denomination	Paper	Ink	Serial	Block	Circ.	Unc.
NE-3-1-1	1 cent	buff	black	red	C	75	150
NE-3-1-5	5 cents	buff	black	red	C	75	150
NE-3-1-10	10 cents	buff	black	red	C	100	200
NE-3-1-25	25 cents	buff	black	red	C	150	300
NE-3-1-50	50 cents	pink	black	red	C	200	400

Ft. Crook

Ft. Crook was located just outside of Omaha. The camp was previously a CCC camp but was opened in 1943 to hold 1,620 Italian POWs. The camp closed in May 1946.

NE-10-1-1a

NE-10-1-1b

No.	Denomination	Paper	Ink	Serial	Block	Circ.	Unc.
NE-10-1-1a	1 cent	pale green	red	red	A	175	350
NE-10-1-1b	1 cent	pale yellow	black	red	BL	175	350

Ft. Robinson

Ft. Robinson was planned to be built in the northwest corner of Nebraska to house 1,000 POWs. The purpose of the camp was to have the POWs take care of thousands of remount mules and horses. The camp was later expanded to hold over 3,000 POWs. The first group of 600 German POWs arrived from Rommel's Afrika Korps on November 19, 1943. The POWs entertained themselves at the movie theater and on the soccer fields. In early 1945, the camp was designated a naval camp and the German army POWs were replaced by German naval personnel. In the Fall of 1945, the camp became a sub-camp of Scottsbluff and finally closed in May 1946.

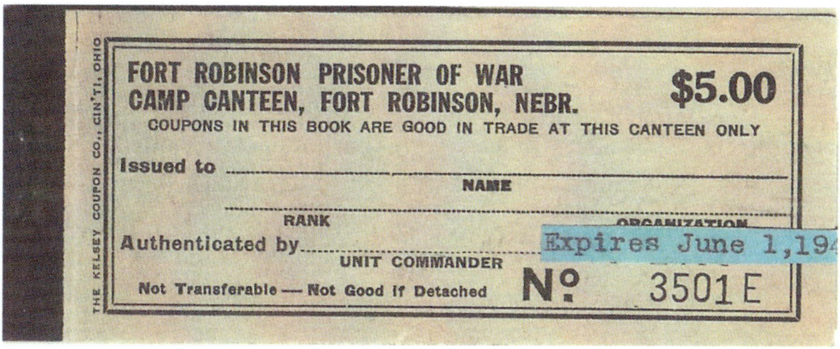

NE-11-1-1

NE-11-1-5

NE-11-1-10

NE-11-1-25

No.	Denomination	Paper	Ink	Serial	Block	Circ.	Unc.
NE-11-1-1	1 cent	gray	blue	red	A,F	25	40
NE-11-1-5	5 cents	pale yellow	blue	red	A,F	35	55
NE-11-1-10	10 cents	pale orange	blue	red	A,F	55	85
NE-11-1-25	25 cents	pale green	blue	red	A,F	90	150

Indianola

Indianola is located in the southwestern portion of Nebraska and was selected in 1942 to be a site for a POW camp. The camp was opened in February 1944 and held 721 POWs. It closed in November 1945.

NE-17-1-1

No.	Denomination	Paper	Ink	Serial	Block	Circ.	Unc.
NE-17-1-1	1 cent	buff	black	red		125	250

Scottsbluff

Camp Scottsbluff was located in Scottsbluff, Nebraska, in the western part of the state near the Wyoming border. The camp opened in February 1944 and closed March 1946. The camp held 2,648 German POWs. Few incidents occurred involving the POWs. However, there was one protest that went on for a few days due to a delay in the next shipment of beer. POWs banged on pots and pans day and night to express their displeasure. They were already irritated because they were sold 3.2% beer, which was referred to as American water beer. The camp was closed in March 1946.

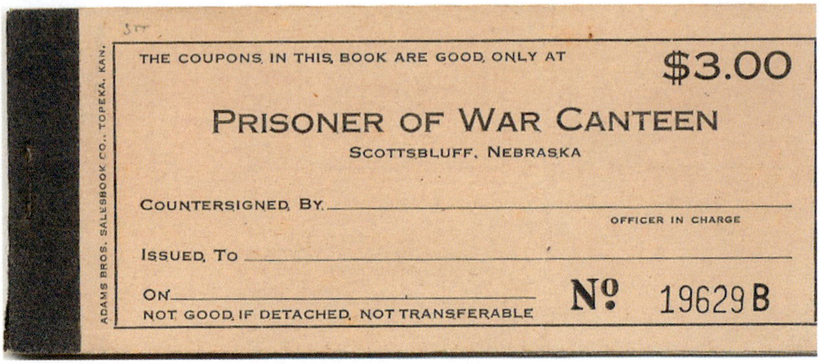

NE-25-1-2

NE-25-1-5

No.	Denomination	Paper	Ink	Serial	Block	Circ.	Unc.
NE-25-1-2	2 cents coupon check	pale orange	blue	red	B	150	300
NE-25-1-5	5 cents coupon check	buff	blue	red	B	150	300

NEVADA

There were no camps located in Nevada.

NEW HAMPSHIRE

There was only one POW camp located in New Hampshire. The camp was located in Stark, NH, in Coos County in the central part of the state. In 1944, an old CCC camp was converted into a POW camp for 250 German POWs, several hundred of whom were transferred from Ft. Devens, MA. They were used in nearby forests to supply wood for paper mills in nearby Berlin, NH. Stark was actually a sub-camp of the Houlton, ME base camp. Stark was closed in the spring of 1946.

	Camp	Base Camp
1	Stark	Houlton, ME

NEW JERSEY

New Jersey had both POW camps and Italian Service Units (ISU) during World War II. Italian Service Units were created after the surrender of Italy. The POWs had more freedom since they were now non-combatants and could perform labor in defense related industry. There were two ISUs, one at Bellemead and the other at Raritan Arsenal. POW base camps were located at Camp Kilmer, Ft. Dix, Ft. Monmouth, Jersey City QMD, and Sommerville QMSD. There were approximately 6,800 POWs held in POW camps and 1,250 Italians held in ISUs in New Jersey.

	Camp	Base Camp
1	*Belle Mead ISB	
2	Bridgeton	Ft. Dix, NJ
3	*Camp Kilmer	
4	Caven Point	Camp Kilmer, NJ
5	Centerton	Ft. Dix, NJ
6	Dias Creek	Ft. Dix, NJ
7	*Fort Dix	
8	Fort Monmoth	
9	Glassboro	Ft. Dix, NJ
10	Jersey City QMD	
11	Paravan State Park	Ft. Dix, NJ
12	*Raritan Arsenal ISB	
13	Somerville QMSD	

Belle Mead ASF

Belle Mead was formally listed as an Italian Service Unit; however, the camp also held POWs. Therefore, the chits labeled ASF are listed with POW.

NJ-1-2-1

NJ-1-2-5

NJ-1-2-10

NJ-1-2-25

NJ-1-2-50

No.	Denomination	Paper	Ink	Serial	Block	Circ.	Unc.
	ASF Depot						
NJ-1-2-1	1 cent	white	blue	red	E	25	40
NJ-1-2-5	5 cents	rose	blue	red	E	35	55
NJ-1-2-10	10 cents	pale orange	blue	red	E	55	85
NJ-1-2-25	25 cents	pale gray	blue	red	E	90	150
NJ-1-2-50	50 cents	green	blue	red	E	125	175

Camp Kilmer

Camp Kilmer was named after the poet Joyce Kilmer and was chosen to be built in 1941 as a staging area for the Port of New York. Construction was completed in June 1942, with the camp being located near Edison, NJ. A small portion of the camp was set aside as a POW camp beginning in May 1945 and closing in July 1946. The camp held both Italian and German POWs. Camp Kilmer was closed as a military base in 2009. The camp actually was used as a POW camp for Germans and an ISU for the Italians.

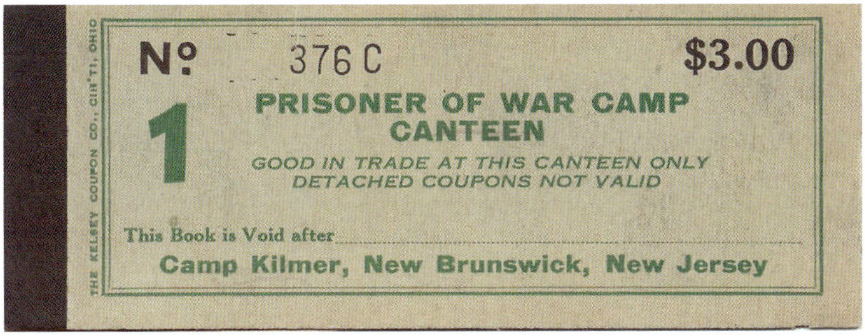

NJ-3-1-1a

NJ-3-1-1b

NJ-3-1-1c

NJ-3-1-5a

NJ-3-1-5b

NJ-3-1-5c

NJ-3-1-10a

NJ-3-1-10b

NJ-3-1-25a

NJ-3-1-25b

No.	Denomination	Paper	Ink	Serial	Block	Circ.	Unc.
	Camp Canteen						
NJ-3-1-1a	1 cent	green	green	red	1 C	25	40
NJ-3-1-1b	1 cent	orange	black	red	1 C	25	40
NJ-3-1-1c	1 cent	green	blue	red	4 C	25	40
NJ-3-1-5a	5 cents	rose	green	rec	1 C	35	55
NJ-3-1-5b	5 cents	pale orange	black	red	1 C	35	55
NJ-3-1-5c	5 cents	rose	blue	red	4 C	35	55
NJ-3-1-10a	10 cents	gray	green	red	1 C	55	85
NJ-3-1-10b	10 cents	pale yellow	black	red	1 C	55	85
NJ-3-1-10c	10 cents	gray	blue	red	4 C	55	85
NJ-3-1-25a	25 cents	pale orange	green	red	1 C	90	150
NJ-3-1-25b	25 cents	buff	black	red	1 C	90	150
NJ-3-1-25c	25 cents	pale orange	blue	red	4 C	90	150

Ft. Dix

Ft. Dix is located 16 miles southeast of Trenton, NJ and was established in 1917 as a staging and training camp for World War I. The base is named after Major General John Dix, who served in the War of 1812 and the Civil War. During World War II, Ft. Dix again served as a staging and training base, and also held 3,000 German and Italian POWs, as well as Russians who fought for the Nazis. The POW camp operated from January 1944 until June 1946. On June 29, 1945, a riot and attempted mass suicide led by 154 Soviet born prisoners broke out. They knew that they were going to be repatriated to the Soviet Union and not Germany. They were part of a contingent of almost 1 million Soviets that fought for Germany. Most were captured in France where they were guarding coastal installations. Despite pleas from US Government officials to allow them to stay in the US, they were sent back to the USSR on August 31, 1945, never to be heard from again.

NJ-7-1-1

NJ-7-1-25

NJ-7-2-5

No.	Denomination	Paper	Ink	Serial	Block	Circ.	Unc.
NJ-7-1-1	1 cent	pale yellow	blue	red	A	150	300
NJ-7-1-25	25 cents	mauve	blue	red	A	200	400
NJ-7-2-5	5 cents	pink	green	red	3 C	200	400

Ft. Monmouth

Ft. Monmouth was an army installation located between the cities of Eatontown, Tinton Falls, and Oceanport, NJ. During World War II it was used as a training base for the signal corps. The base served as a POW camp from November 1945 until May 1946, housing 800 POWs.

Jersey City Quartermaster Supply Depot

Not much is known about this facility other than it contained a POW camp from March 1944 until September 1945 and held 300 POWs.

Somerville Quartermaster Supply Depot

The Somerville Depot was constructed in 1942 and 1943 as an Army Quartermaster Depot and POW camp. The POW camp operated from January 1945 until May 1946 and housed 200 POWs.

NEW MEXICO

New Mexico was home to 18 POW camps, with base camps at Lordsburg, Roswell, and Ft. Bliss Texas. Camps in New Mexico held German and Italian POWs, as well as Japanese internees. Over 10,000 POWs were located in New Mexico during World War II, helping local farmers and migrant Mexican workers harvest crops.

	Camp	Base Camp
1	Alamagordo	Fort Bliss, TX
2	Albuquerque	Roswell, NM
3	Anthony	Fort Bliss, TX
4	Artesia	Roswell, NM
5	Bruns General Hospi-	Roswell, NM
6	Carlsbad	Roswell, NM

7	Clayton	Roswell, NM
8	Clovis AAF	Roswell, NM
9	Deming	Lordsburg, NM
10	Dexter	Roswell, NM
11	Fort Bayard	Lordsburg, NM
12	Fort Sumner	Roswell, NM
13	Hatch	Fort Bliss, TX
14	Kirtland Field	Roswell, NM
15	Las Cruces	Fort Bliss, TX
16	*Lordsburg	
17	Mayhill	Roswell, NM
18	Melrose	Roswell, NM
19	Portales	Roswell, NM
20	*Roswell	

Lordsburg

Lordsburg ,New Mexico is located in the extreme southwestern corner of the state. The area during World War II contained POW camps that held German and Italian prisoners and a Japanese Internment camp. Construction of the camp began in February 1942. In July 1942, 630 Japanese American Issei men were transferred from Ft. Lincoln Internment camp in Bismarck, North Dakota. Eventually 1,500 Japanese were interred at the camp. The camp was run by the Department of Justice. All of the Japanese Americans were removed from the camp by July 1943, and 4,000 Italian and German POWs were brought in to replace them. The camp was closed in June 1945.

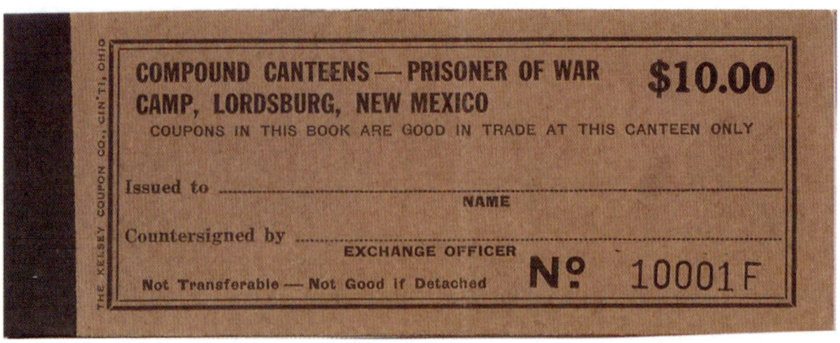

THE KELSEY COUPON CO., CIN'TI, OHIO

COMPOUND CANTEENS — PRISONER OF WAR CAMP, LORDSBURG, NEW MEXICO $10.00

COUPONS IN THIS BOOK ARE GOOD IN TRADE AT THIS CANTEEN ONLY

Issued to .... NAME

Countersigned by .... EXCHANGE OFFICER

Not Transferable — Not Good if Detached Nº 10001 F

NM-16-1-1a

NM-16-1-1b

NM-16-2-1

NM-16-2-5

NM-16-2-10

NM-16-2-25

NM-16-2-50

No.	Denomination	Paper	Ink	Serial	Block	Circ.	Unc.
	Compounds Canteen						
NM-16-1-1a	1 cent	buff	black	red		150	300
	Compounds Canteen 3-26-45						
NM-16-1-1b	1 cent	red	black	black		150	300
	Denomination on right						
NM-16-2-1	1 cent	pale orange	blue	red	B,E,F	25	40
NM-16-2-5	5 cents	buff	blue	red	B,E,F	35	55
NM-16-2-10	10 cents	red	blue	blue	E,F	55	85
NM-16-2-25	25 cents	rose	blue	red	E,F	90	150
NM-16-2-50	50 cents	orange	blue	red	F	125	175

Roswell

Roswell is located in southeastern New Mexico, and a POW camp was built in Orchard Park, an area 14 miles southwest of town. The first German POWs from Rommel's Afrika Korps arrived in January 1, 1943. A total of 4,000 German POWs were housed in the camp, which covered 120 acres. The camp had a fully equipped hospital, a library, garden, and a newspaper. The primary work for the POWs was picking cotton in area fields. A major project accomplished by the POWs was shoring up the North Spring River in Roswell. An iron cross fashioned out of rocks by the POWs can still be seen on one of the banks. Local farmers donated food to the camp to supplement the diet of the POWs, which at one time was cut to 800 calories. This greatly improved the attitude of the German POWs toward Americans. The camp closed in January 1946.

NM-20-1-1

NM-20-1-5

NM-20-1-25

NM-20-2-1

NM-20-2-5

NM-20-2-10

NM-20-2-25a

NM-20-2-25b

No.	Denomination	Paper	Ink	Serial	Block	Circ.	Unc.
	Camp Canteen						
NM-20-1-1	1 cent	buff	black	red	E	125	250
NM-20-1-5	5 cents	pink	red	black	P	150	300
NM-20-1-25	25 cents	gray	red	black	P	150	300
	Exchange						
NM-20-2-1	1 cent	orange	blue	red	C,D	25	40
NM-20-2-5	5 cents	rose	blue	red	C,D	35	55
NM-20-2-10	10 cents	buff	blue	red	C,D	55	85
NM-20-2-25a	25 cents	tan	blue	red	C	90	150
NM-20-2-25b	25 cents	pale yellow	blue	red	D	90	150

NEW YORK

New York was home to both POW camps and Italian Service Units. There were a total of 48 POW camps, with base camps at Ft. Niagara and Pine Camp; and 5 ISUs, the main one at Raritan Arsenal. Other references include Elmira ASF, Halloran, Hamilton, Jay, Mitchel Field, Sampson, Shanks, Upton, Van Etten, and West Point as base camps. An alien internment camp that held Germans, Italians, and Japanese was located at Ellis Island. The first POW camp in New York was opened in 1943 and was located at Pine Camp, now Ft. Drum. Due to the rising numbers of POWs, many were moved to branch camps in western New York. Because of the increased distance from Pine Camp, a new base camp was established in Ft. Niagara in June 1944.

	Camp	Base Camp
1	Attica	Niagara Falls, NY
2	Boonville	Pine Camp, NY
3	Brocton	Niagara Falls, NY
4	Clyde	Pine Camp, NY
5	Cobbs Hill	Niagara Falls, NY
6	Dunkirk	Niagara Falls, NY
7	Elmira ASF Depot	Pine Camp, NY
8	Fair Haven	Pine Camp, NY
9	Fayetteville	Pine Camp, NY
10	Fort Hamilton	
11	Fort Jay	

12	*Fort Niagara	
13	Genesee	Niagara Falls, NY
14	Geneva	Pine Camp, NY
15	Green Haven USDB	
16	*Halloran General Hospital Staten Island	
17	Hamlin Beach	Niagara Falls, NY
18	Harrisville	Pine Camp, NY
19	Letchworth Park	Niagara Falls, NY
20	Mason General Hospital	
21	Marion	Niagara Falls, NY
22	Medina	Niagara Falls, NY
23	*Mitchell Field	
24	Naples	Niagara Falls, NY
25	Newark	Niagara Falls, NY
26	*New York Port of Entry	
27	Oakfield	Pine Camp, NY
28	Ontario	Pine Camp, NY
29	*Pine Camp 1209th SCU,	1277th SCSU
30	Popolopen (West Point)	
31	Port Bryon	Pine Camp, NY
32	Rochester	Pine Camp, NY
33	Rome	Pine Camp, NY
34	*Sampson 1215th SCU	
35	Shanks Staging Area	
36	*Shanks	
37	Sodus Point	Niagara Falls, NY
38	Slaterville Springs	Pine Camp, NY
39	Upton	
40	Van Etten	
41	Voorheesville	Pine Camp, NY
42	Wadsworth	
43	Westfield	Niagara Falls, NY

Ft. Niagara

Ft. Niagara opened as a POW camp on June 12, 1944 and was designed to hold up to 3,000 POWs. The camp consisted of 54 buildings and had its own school, orchestra, and choir in addition to sports activities. Most of the POWs worked on local farms and in food processing plants. Ft. Niagara was the base camp for 13 branch camps in western and central New York. In October 1944, a top secret branch was located in Van Etten between Ithaca and Elmira and was nicknamed the idea factory. It consisted of US intelligence specialists and German POWs opposed to Hitler. The operation was very successful and eventually was moved to Ft. Kearny, RI but still used Ft. Niagara as its address. Ft. Niagara began to release prisoners in late 1945 and closed in January 1946.

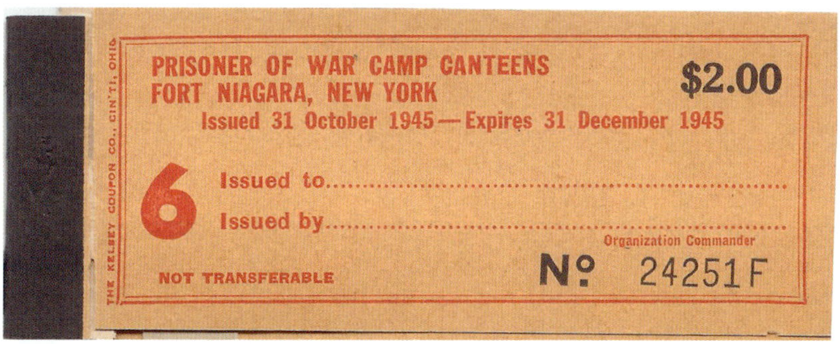

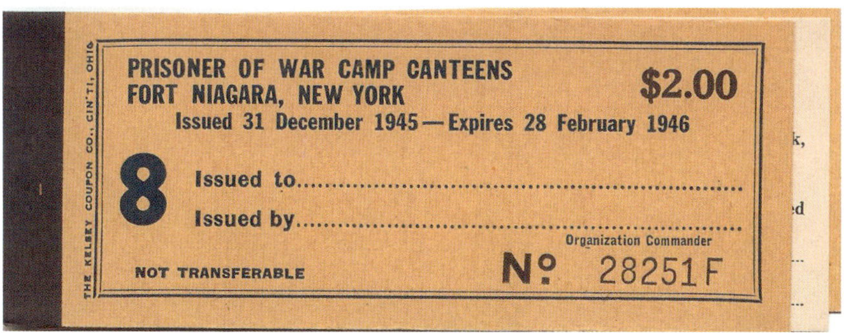

NY-12-1-1

NY-12-2-1a

NY-12-2-1b

NY-12-2-1c

NY-12-2-5a

NY-12-2-5b

NY-12-2-5c

NY-12-2-10a

NY-12-2-10b

NY-12-2-10c

No.	Denomination	Paper	Ink	Serial	Block	Circ.	Unc.
	PW Camp						
NY-12-1-1	1 cent	white	blue	red	B	100	200
	PW						
NY-12-2-1a	1 cent	pale orange	blue	red	F	25	40
NY-12-2-1b	1 cent	pale orange	green	red	F	25	40
NY-12-2-1c	1 cent	pale orange	red	red	F	25	40
NY-12-2-5a	5 cents	buff	blue	red	F	35	55
NY-12-2-5b	5 cents	buff	green	red	F	35	55
NY-12-2-5c	5 cents	pale yellow	red	red	F	35	55
NY-12-2-10a	10 cents	buff	blue	red	F	55	85

NY-12-2-10b	10 cents	pale gray	blue	red	F	55	85
NY-12-2-10c	10 cents	pale gray	green	red	F	55	85
NY-12-2-10d	10 cents	gray	red	red	F	55	85

Pine Camp

Pine Camp was first established in 1908 to train northeastern National Guard and Reserve units. The camp was located in Jefferson County, NY. Beginning with the outbreak of World War II the camp was expanded by 75,000 acres. In 1943, Italian POWs began to arrive and were later followed by German POWs. There were approximately 2,500 Italian POWs and 1,500 German POWs. The POWs were employed as lumberjacks and farm hands. The POW camp closed in May 1946 and Pine Camp was renamed Camp Drum in 1951 and Fort Drum in 1974.

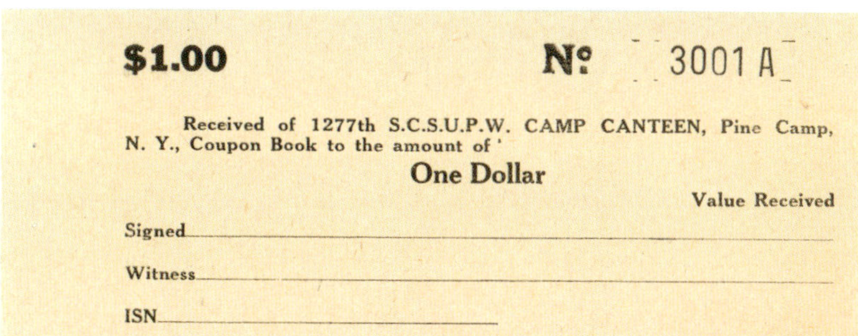

NY-29-1-1

NY-29-1-5

NY-29-2-5

NY-29-2-10

No.	Denomination	Paper	Ink	Serial	Block	Circ.	Unc.
NY-29-1-1	1 cent S.C.S.U.	green	blue	red	A	100	200
NY-29-1-5	5 cents S.C.S.U.	orange	blue	red	A	125	250
NY-29-2-5	5 cents S.C.U.	orange	blue	red	H	125	250
NY-29-2-10	10 cents S.C.U.	pale green	blue	red	A	125	250

Halloran General Hospital

Halloran General Hospital is located on Staten Island.

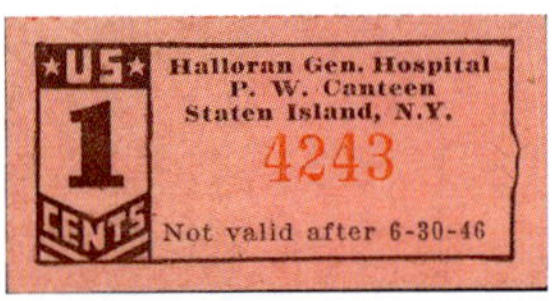

NY-16-1-1

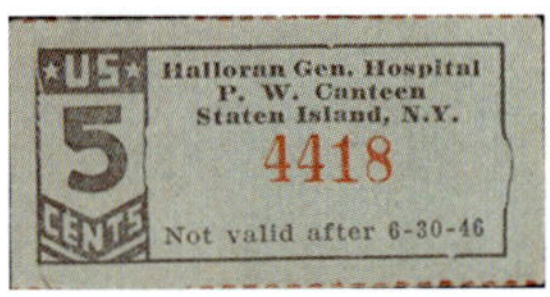

NY-16-1-5

NY-16-1-25

No.	Denomination	Paper	Ink	Serial	Block	Circ.	Unc.
NY-16-1-1	1 cent not valid after 6-30-46	pink	black	red		175	350
NY-16-1-5	5 cents not valid after 6-30-46	pale green	black	red		175	350
NY-16-1-25	25 cents not valid after 6-30-46	buff	black	red		200	400

Sampson

NY-34-1-1

No.	Denomination	Paper	Ink	Serial	Block	Circ.	Unc.
NY-34-1-1	1 cent	pale orange	blue	red		150	300

German prisoners boarding trucks on July 22, 1945 at Camp Shanks, New York for the trip to the ship that will return them to their homeland. (AP wirephoto)

Italian POWs helping harvest grapes on the William Russo farm near Freedonia, New York in 1943. The pay was 80 cents a day, far more than the 26 cents they got in the Italian army. Each prisoner received a credit of $3.00 per month from his earnings for use at the camp canteen. The balance was given to him when he was freed. (ACME)

Image courtesy of the State Archives, North Carolina Office of Archives and History.

Crew members of the German U-352, captured in May 1942 off the North Carolina coast after the U.S. Coast Guard sank their vessel, became prisoners of war at For t Bragg.

©2008 North Carolina Museum of History
Office of Archives and History, N.C. Department of Cultural Resources

Prisoners often made items to trade, sell, or give away. This watercolor landscape of the barracks at Camp Butner was painted in 1945 and signed "E Kamler." It came to the North Carolina Museum of History by way of the niece of a guard at that camp, who may have purchased it or received it as a gift.

NORTH CAROLINA

The first POWs of the war were brought to Ft. Bragg in the spring of 1942 when the U-Boat U-352 was torpedoed and sunk. Thirteen German sailors died, but 33 survived and became POWs. There were 18 camps in North Carolina, three of which were base camps. The base camps were Butner, Ft. Bragg, and Mackall. Additionally, there were base camps in Aliceville, Alabama and Forrest, Tennessee.

	Camp	Base Camp
1	Ahoskie	Camp Butner, NC
2	*Camp Butner	
3	Camp Davis Xerox Fantasy	Camp Butner, NC
4	*Camp Mackall	
5	* Fort Bragg	
6	Greensboro	Camp Butner, NC
7	Hendersonville	Camp Forrest, TN
8	Seymour Johnson Field	Camp Butner, NC
9	Monroe	Camp Butner, NC
10	Moore General Hospital	Camp Butner, NC
11	New Bern	Camp Butner, NC
12	Scotland Neck	Ft. Bragg, NC
13	Sutton	Aliceville, AL
14	Whiteville	Camp Butner, NC
15	Williamston	Camp Butner, NC
16	Wilmington	Camp Butner, NC
17	Winston-Salem	Camp Butner, NC

Camp Butner

Camp Butner was located in Butner, North Carolina in Granville County and was named after Army General Henry W. Butner. It was used as a staging area during World War II and was also the location of a Battalion Surgeons Assistant school. Camp Butner became the home to 3,000 Italian POWs beginning in 1943. After the surrender of Italy some of these POWs were sent to an ISU camp in Sutton, North Carolina. Later German POWs began to arrive with 3,000 eventually held at the camp. Not all of the POWs were actually German. Camp Butner's "nationalities" compound at one point held 332 Czechs, 150 Poles, 147 Dutchmen, 117 Frenchmen, 34 Austrians, 11 Luxembourgers, and 1 Lithuanian basically forced to fight for the German army. The POW camp closed in May 1946, and Camp Butner closed January 31, 1947.

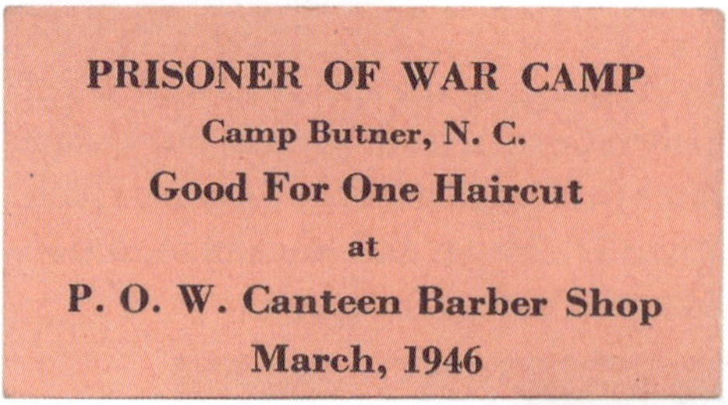

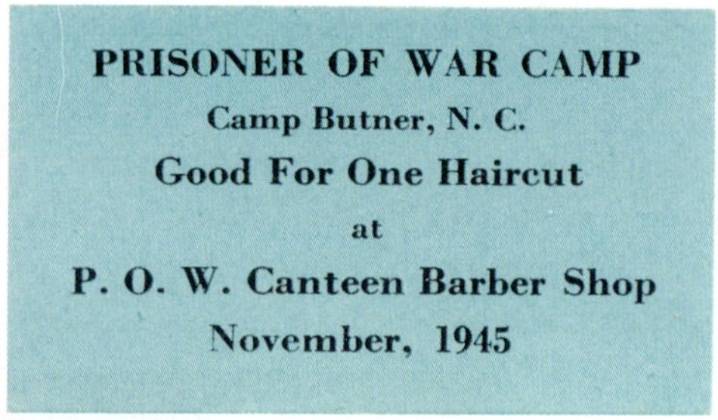

NC-2-1-1

No.	Denomination	Paper	Ink	Serial	Block	Circ.	Unc.
NC-2-1-1	1 cent Red 1	gray	blue	red	D	150	300

Ft. Bragg

Camp Bragg was completed on September 4, 1918 and was used to train Army Artillery Units. On September 30, 1922, Camp Bragg became Ft. Bragg. A massive expansion took place in 1941, with the fort eventually housing 150,000 men. A small number of German POWs (U-352) arrived at the camp in 1942 but were soon transferred out. The main body of German POWs arrived in May 1944. The breakdown of the German POWs was 55% Afrika Korps and 45% captured in France. The POWs primarily performed farm labor picking cotton and tobacco. They also harvested pulp wood and assisted in mosquito control. The POWs began leaving the camp in Fall 1945, and the camp was closed in April 1946, holding over 2,000 POWs.

NC-5-1-1a

NC-5-1-1b

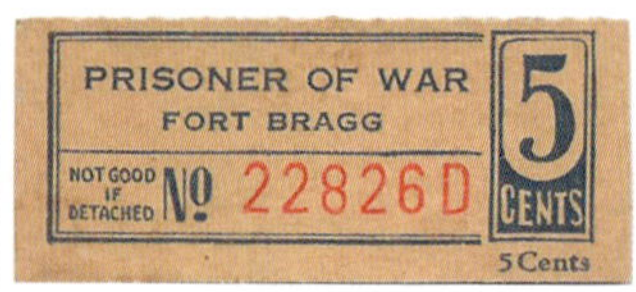

NC-5-1-5

NC-5-1-10

No.	Denomination	Paper	Ink	Serial	Block	Circ.	Unc.
NC-5-1-1a	1 cent	beige	blue	red	D	150	300

NC-5-1-1b	1 cent	gray	blue	red	B	150	300
NC-5-1-5	5 cents	beige	blue	red	D	150	300
NC-5-1-10	10 cents	beige	blue	red	D	150	300

Camp Mackall

Camp Mackall is located in Eastern Richmond County, North Carolina and was known as Camp Hoffman until February 8, 1943. The camp was used to train paratroopers during World War II. Today it is used to train Army Special Forces soldiers. In December 1944, 400 German POWs were housed in the camp. The POW section of the camp closed in January 1946.

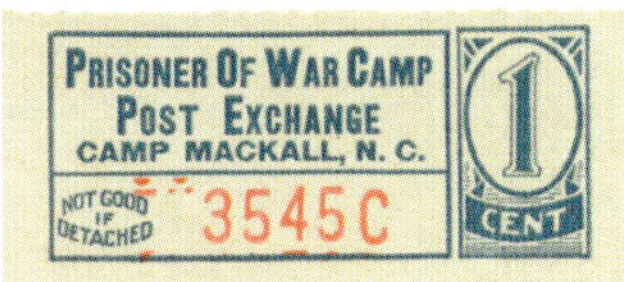

NC-4-1-1

NC-4-1-2

NC-4-1-5a

NC-4-1-5b

NC-4-1-10

NC-4-1-25

No.	Denomination	Paper	Ink	Serial	Block	Circ.	Unc.
NC-4-1-1	1 cent	white	blue	red	B	150	300
NC-4-1-2	2 cents	pale tan	blue	red	B	125	250
NC-4-1-5a	5 cents	pale mauve	blue	black	B	150	300
NC-4-1-5b	5 cents	pale blue	blue	red	D	150	300
NC-4-1-10	10 cents	orange	blue	red	C	150	300
NC-4-1-25	25 cents	blue	blue	red	D	175	350

Camp Davis

All of the chits that have been seen from this camp are fake. They are seen in various denominations, are larger than the average chit, and lack perforations. An example is shown below.

NORTH DAKOTA

There were no base camps located in North Dakota. However, there was a sub-camp at Hazen. This camp operated out of a base camp in Scottsbluff, Nebraska. There are also reports of POWs being sent from Minnesota to work on projects in North Dakota. There was a alien internment camp at Ft. Lincoln that housed German Internees.

	Camp	Base Camp
1	Hazen	Scottsbluff, NE

OHIO

	Camp	Base Camp
1	Bowling Green	Camp Perry, OH
2	*Camp Perry	
3	Celina	Camp Perry, OH
4	Columbus ASF Depot	Camp Perry, OH
5	Crile General Hospital	Camp Perry, OH
6	Defiance	Camp Perry, OH
7	Fletcher General Hospital	Camp Perry, OH
8	Marion Engineer Depot	Camp Perry, OH
9	Scioto Ordinance Depot	Camp Perry, OH
10	Willmington	Camp Perry, OH
11	Wright Field	Atterbury, IN

Camp Perry

This POW camp was located in the western part of Port Clinton, Ohio, at the western part of Lake Erie. This was a base camp for nine sub-camps. The 1970 book on USA POW Camps by Albert Donn did not have any chits listed. Since that book was published, chits have been found. The camp is still in use, primarily by the Ohio National Guard. It is also the site of the National Shooting matches.

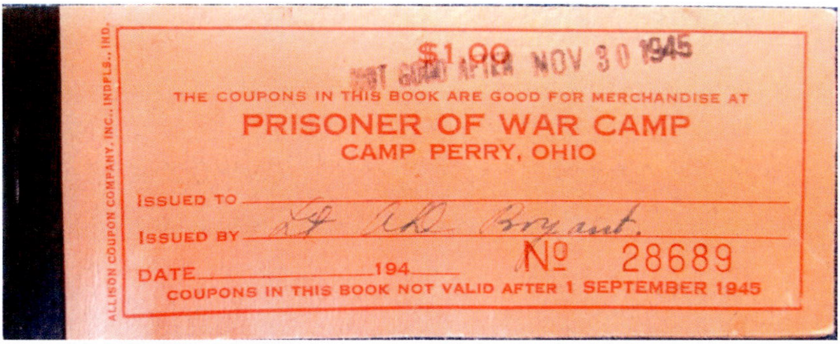

OH-2-1-1b

OH-2-1-10

OH-2-1-25

No.	Denomination	Paper	Ink	Serial	Block	Circ.	Unc.
OH-2-1-1a	1 cent	pale green	red	red		125	250
OH-2-1-1b	1 cent	tan	blue	red	A	150	300
OH-2-1-5	5 cents	pale yellow	red	red		150	300
OH-2-1-10	10 cents	orange	red	red		150	300
OH-2-1-25	25 cents	mauve	red	red		200	400

OKLAHOMA

	Camp	Base Camp
1	*Alva	
2	Ardmore AAB	Howze, TX
3	Bixby	Camp Gruber, OK
4	*Camp Gruber	
5	Chickasha	Ft. Reno, OK
6	Cikasa	Alva, OK
7	*Fort Reno (El Reno)	
8	*Fort Sill	
9	*Glennan General Hospital	
10	Haskell	Camp Gruber, OK

11	Hobart	Ft. Sill, OK
12	*Madill	
13	*McAlester	
14	Okemah	Camp Gruber, OK
15	Okmulgee	Camp Gruber, OK
16	Pauls Valley	Camp Chaffee, AR
17	Porter	Camp Gruber, OK
18	Powell	Howze, TX
19	*Pryor	
20	Will Rogers Field	Ft. Reno, OK
21	Springtown	McAlester, OK
22	Tipton	Ft. Sill, OK
23	Tishomingo	Howze, TX
24	*Tonkawa	
25	Waynoka	Alva, OK
26	Wetumka	Camp Gruber, OK

Alva

This camp was located adjacent to the city of Alva, Oklahoma. The prime farmland, over 650 acres, was donated by two local farmers. When the camp closed in November 1945, the land was given to the city of Alva for an airport, and the farmers didn't get anything in return. The camp opened in August 1943 to primarily house regular army German troops. The records show there were three sub-camps associated with Alva.

Kelsey Coupon Company was the primary supplier of POW chit booklets.

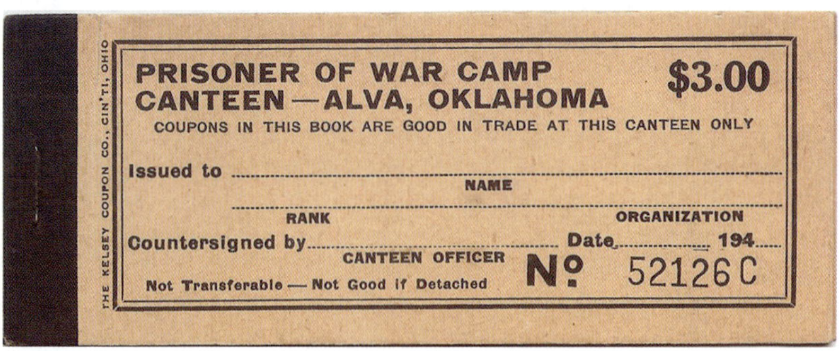

THE KELSEY COUPON CO., CIN'TI, OHIO

PRISONER OF WAR CAMP CANTEEN—ALVA, OKLAHOMA $3.00

COUPONS IN THIS BOOK ARE GOOD IN TRADE AT THIS CANTEEN ONLY

Issued to .... NAME

.... RANK ORGANIZATION

Countersigned by........ CANTEEN OFFICER Date........ 194....

Not Transferable — Not Good if Detached

Nº 52126 C

OK-1-1-1a

OK-1-1-1b

OK-1-1-5

OK-1-1-10a

OK-1-1-10b

OK-1-1-25a

OK-1-1-25b

OK-1-1-25c

No.	Denomination	Paper	Ink	Serial	Block	Circ.	Unc.
OK-1-1-1a	1 cent	pale green	blue	red	C	25	40
OK-1-1-1b	1 cent	white	blue	red	E	25	40
OK-1-1-5	5 cents	pale orange	blue	red	C	35	55
OK-1-1-10a	10 cents	rose	blue	red	C	55	85
OK-1-1-10b	10 cents	tan	blue	red	E	55	85
OK-1-1-25a	25 cents	tan	blue	red	C	90	150
OK-1-1-25b	25 cents	pale orange	blue	red	C,E	90	150
OK-1-1-25c	25 cents	orange	blue	red	E	90	150

Ft. Reno

The Fort Reno POW camp was located on 94 acres of land in the eastern portion of the Fort Reno military base. It opened in June of 1943. The camp held over 1,300 German POWs, mostly from General Romel's Africakorp. They were hired out to work as laborers for local farmers. The POWs constructed a chapel on the on the north side of the parade grounds. The camp closed in June 1946.

Chits were not known to exist until one without a name was discovered. On the back someone wrote "El Reno."

OK-7-1-1

No.	Denomination	Paper	Ink	Serial	Block	Circ.	Unc.
OK-7-1-1	1 cent	gray	blue	red		150	300

Ft. Sill

The POW camp was located on the west side of the military reservation. The camp opened in February 1944 as a base camp, with a maximum capacity of 2,965. The largest number of POWs held was 1,834 in July 1945. There were a total of three escapes.

OK-8-1-1

No.	Denomination	Paper	Ink	Serial	Block	Circ.	Unc.
OK-8-1-1	1 cent	buff	blue	red		150	300

Glennan General Hospital

The title of the POW camp located at Okmulgee, Oklahoma was Glennan General Hospital. It was located on the east side of town, east of Mission Road and south of Industrial Drive. The hospital treated American soldiers until October 1944, when it was converted to handle the treatment of POWs with serious injuries. In November 1945 it reverted back to treating American servicemen. The population of POWs was 2,236.

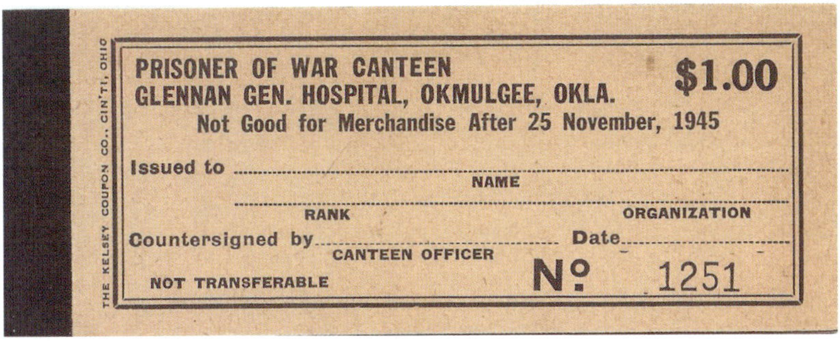

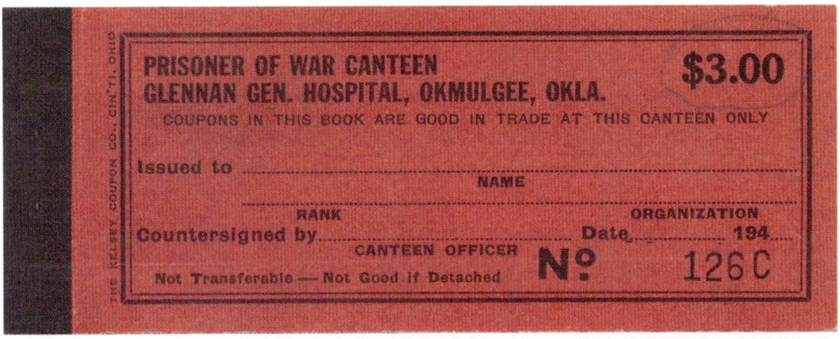

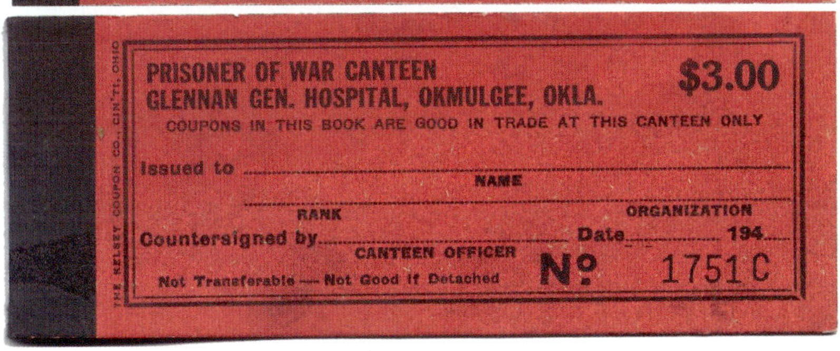

OK-9-1-1a

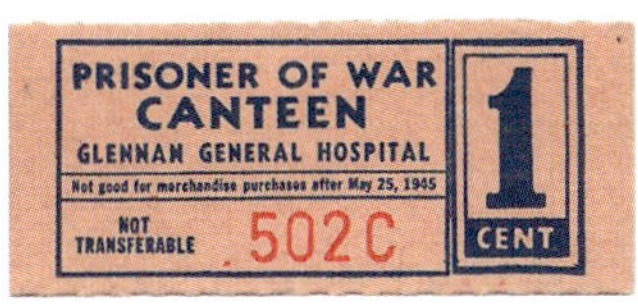

OK-9-1-1b

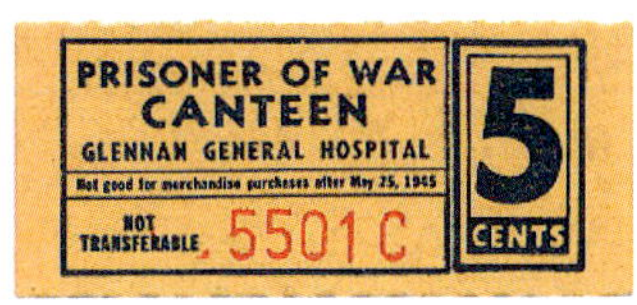

OK-9-1-5b

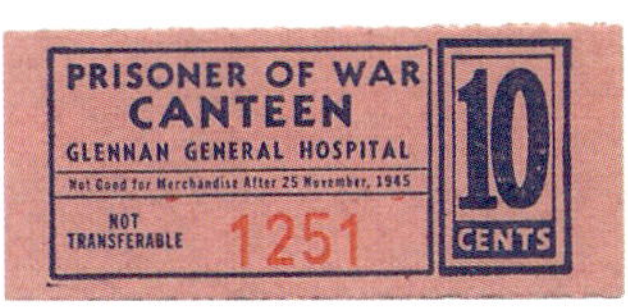

OK-9-1-10a

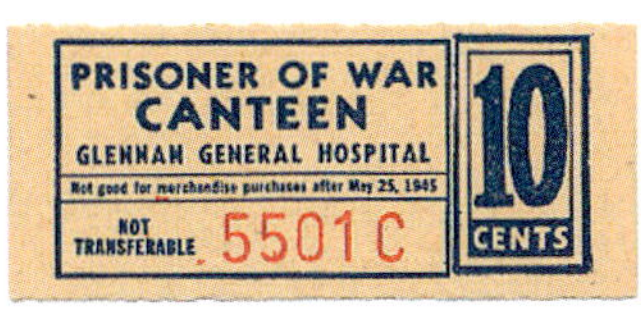

OK-9-1-10b

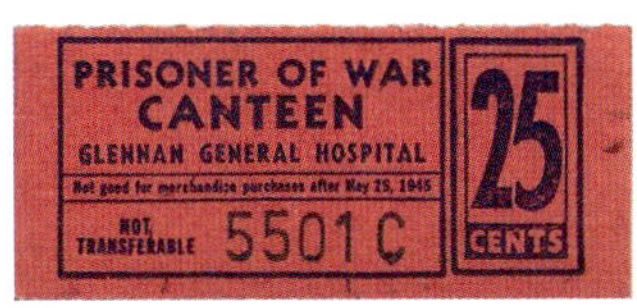

OK-9-1-25

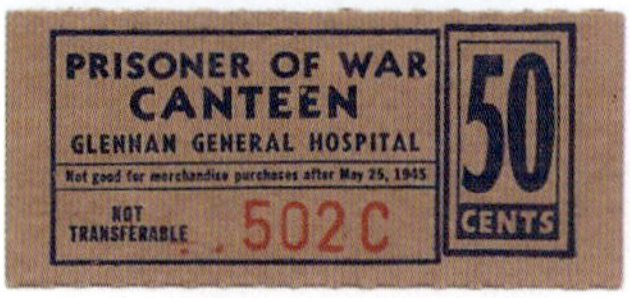

OK-9-1-50a

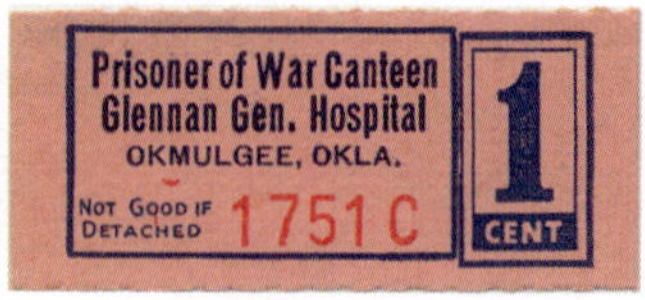

OK-9-2-1

OK-9-2-5

OK-9-2-10

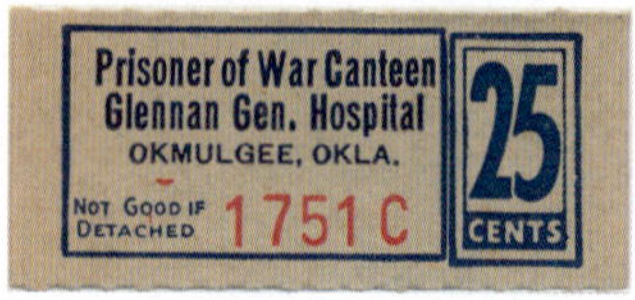

OK-9-2-25

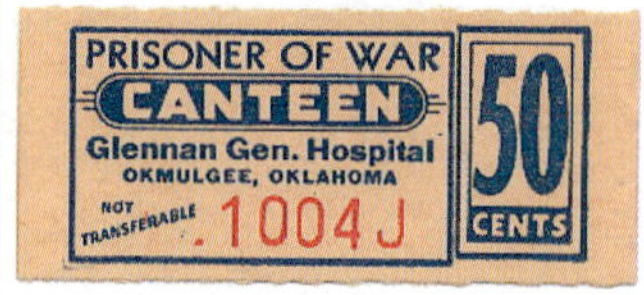

OK-9-3-50

No.	Denomination	Paper	Ink	Serial	Block	Circ.	Unc.
	25 Nov 1945						
OK-9-1-1a	1 cent	orange	blue	red		25	40
OK-9-1-1b	1 cent	pale pink	blue	red	C	25	40
OK-9-1-5a	5 cents	buff	blue	red		35	55
OK-9-1-5b	5 cents	yellow	blue	red	C	35	55
OK-9-1-10a	10 cents	rose	blue	red		55	85
OK-9-1-10b	10 cents	buff	blue	red	C	55	85
OK-9-1-25	25 cents	red	blue	red	C	90	150
OK-9-1-50a	50 cents	brown	blue	red	C	125	200
OK-9-1-50b	50 cents	white	blue	red	C	125	200
	Okmukgee						
OK-9-2-1	1 cent	pink	blue	red	C	25	40
OK-9-2-5	5 cents	pale gray	blue	red	C	35	55
OK-9-2-10	10 cents	orange	blue	red	C	55	85
OK-9-2-25	25 cent	pale green	blue	red	C	90	150
	CANTEEN						
OK-9-3-1	1 cent	pink	blue	red	J	25	40
OK-9-3-5	5 cents	pale tan	blue	red	J	35	55
OK-9-3-25	25 cents	pale gray	blue	red	J	90	150
OK-9-3-50	50 cents	buff	blue	red	J	125	200

Camp Gruber

The POW camp was located across the road from Camp Gruber, on the west side of highway 10. The first reported POWs were on May 31, 1943. The camp closed in April 1946 as the last POW camp in the state of Oklahoma. This base camp had a capacity of 5,750 POWs, but the largest number at one time was 4,702 in October 1945.

OK-4-1-1

OK-4-1-2

OK-4-2-1

No.	Denomination	Paper	Ink	Serial	Block	Circ.	Unc.
OK-4-1-1	1 cent	buff	black	red		100	200
OK-4-1-2	2 cents	gray	black	red		125	250

Camp Madill

In Albert Donn's book, published in 1970, this was listed as a base camp and had two sub-camp chits illustrated. The style of the paper chit is of the very early design.

A check of contemporary internet sources shows it was "located in the Old First National Bank Building in Madill, this camp opened on April 29, 1943, and closed on April 1, 1944. It was not an actual PW camp, but was the administrative headquarters for several camps in the area, including the ones at Powell and Tishomingo. There were no POWs confined there."

OK-12-1-1

OK-12-2-1

No.	Denomination	Paper	Ink	Serial	Block	Circ.	Unc.
	Sub Camp 1						
OK-12-1-1	1 cent	pink	black	red		150	300
	Sub Camp 2						
OK-12-2-1	1 cent	tan	black	red		150	300

McAlester

This base camp opened in June 1943 and closed in October 1945. It was located north of the city. As a base camp it had a capacity of 4,920 but never held more than 2,876 POWs. Later in 1945, it held patients recovering from treatment at Glennan General Hospital POW Camp.

OK-13-1-1

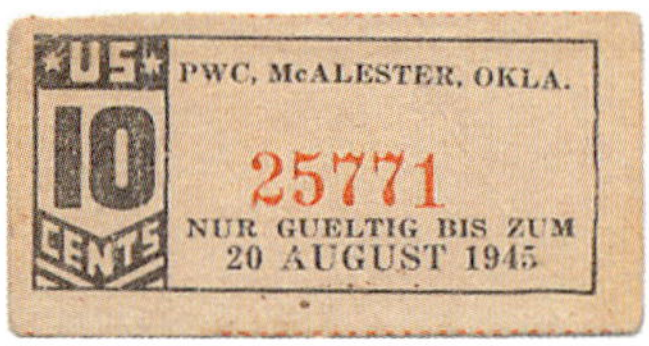

OK-13-1-10

OK-13-2-1

OK-13-2-5

OK-13-2-10

OK-13-2-25

No.	Denomination	Paper	Ink	Serial	Block	Circ.	Unc.
	PWC						
OK-13-1-1	1 cent	gray	black	red		100	200
OK-13-1-10	10 cents	gray	black	red		150	300
	Trade Coupon						
OK-13-2-1	1 cent	pink	red & black	red		100	200
OK-13-2-5	5 cents	pale yellow	red & black	red		125	250
OK-13-2-10	10 cents	pale green	red & black	red		150	300
OK-13-2-25	25 cents	pale rose	red & black	red		175	350

Pryor

The POW Camp was located south of the city of Pryor, east of Highway 69. It opened as a base Camp in November 1944 and closed a short five months later in March 1945. This short operating time might be attributed to the combination of two successful escapes and the close proximity of an explosives plant.

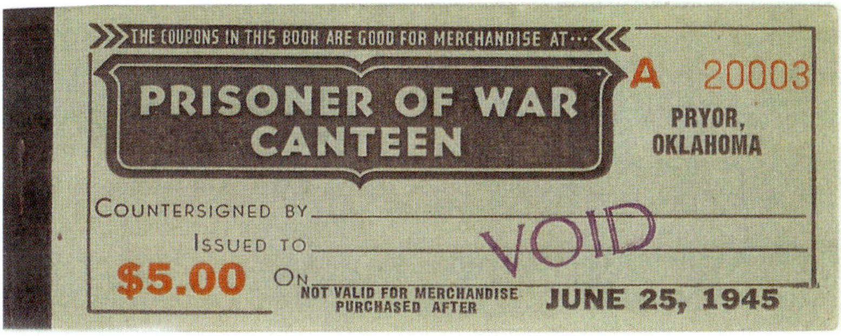

OK-19-1-1

OK-19-1-5

OK-19-1-10

OK-19-1-25

No.	Denomination	Paper	Ink	Serial	Block	Circ.	Unc.
OK-19-1-1	1 cent	buff	black	red	A	100	200
OK-19-1-2	2 cents	gray	black	red	A	175	350
OK-19-1-5	5 cents	pale buff	black	red	A	150	300
OK-19-1-10	10 cents	pale green	black	red	A	175	350
OK-19-1-25	25 cents	orange	black	red	A	200	400

Tonkawa

This 160 acre camp was open for German POWs from January 1943 until September 1945. It held up to 3,000 POWs from the Afrika Korps. There was an escape of 8 POWs in November 1943. They were recaptured shortly after their brief moment of freedom.

OK-24-1-1

OK-24-1-5

No.	Denomination	Paper	Ink	Serial	Block	Circ.	Unc.
OK-24-1-1	1 cent Trade Coupon	pink	red & black	red		150	300
OK-24-1-5	5 cents Trade Coupon	pale yellow	red & black	red		150	300

OREGON

	Camp	Base Camp
1	Adair	
2	*Camp White	
3	Nyassa	Rupert, ID
4	Squaw Creek	Ft. Lewis, WA
5	Vale	Rupert, ID

Camp Adair

This camp was primarily a training camp for troops in World War II. The POW camp for German and Italian prisoners opened in October 1944 and closed in May of 1946. This POW camp held a total of 1,308 POWs.

OR-1-1-1

No.	Denomination	Paper	Ink	Serial	Block	Circ.	Unc.
OR-1-1-1	1 cent red PW	pale orange	black/red	black	A	150	300

Camp White

This Camp was dedicated on September 15, 1942 in Jackson County, Oregon as a training camp. A POW camp was set up on May 1945. A total of 1,156 POWS were encamped there until May of 1946.

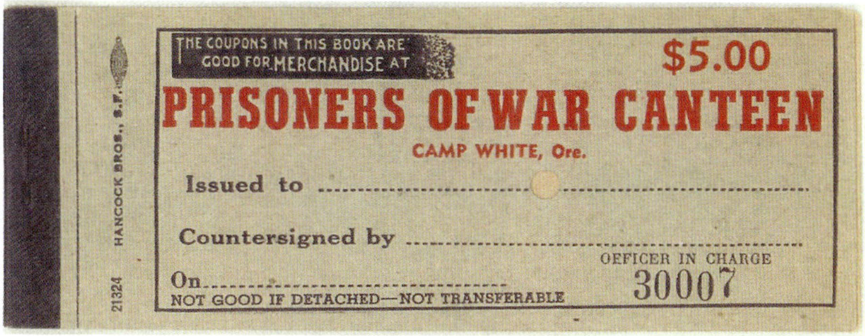

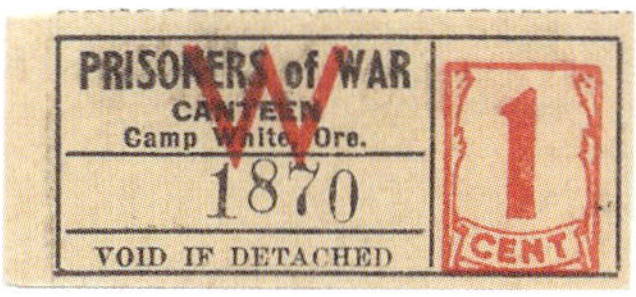

OR-2-1-1

OR-2-1-5

OR-2-1-10

OR-2-1-25

No.	Denomination	Paper	Ink	Serial	Block	Circ.	Unc.
OR-2-1-1	1 cent	pale orange	red & black	black		25	40
OR-2-1-5	5 cents	pink	red & black	black		35	55
OR-2-1-10	10 cents	buff	red & black	black		55	85
OR-2-1-25	25 cents	blue	red & black	black		90	150

PENNSYLVANIA

	Camp	Base Camp
1	Bark Shanty	Indiantown Gap, PA
2	Bull Hill	Camp Reynolds, PA
3	*Camp Reynolds	
4	Carlisle Baracks	
5	Galeton	Indiantown Gap, PA
6	Gettysburg	Indiantown Gap, PA
7	*Indiantown Gap	

8	Johnstown	Indiantown Gap, PA
9	Kane	Camp Reynolds, PA
10	Letterkenny Ordinance Depot	
11	Lewisburg USP	
12	Marionville	Camp Reynolds, PA
13	Mechanicaburg	Indiantown Gap, PA
14	Model Farms	
15	National Guard Armory	Indiantown Gap, PA
16	New Cumberland	
17	North East	Camp Reynolds
18	*Olmstead Field	
19	Philadelphia Naval Yard	Indiantown Gap, PA
20	Reading AAF	Indiantown Gap, PA
21	Sheffield	Camp Reynolds, PA
22	Sideling Hill	Indiantown Gap, PA
23	Stewartstown	Indiantown Gap, PA
24	*Tobyhanna Military Reservation	
25	*Valley Forge General Hospital	

Indiantown Gap

This Lancaster County facility housed German POWS from October 10, 1944 to May 1945. A total of 1,260 were held there.

PA-7-1-5

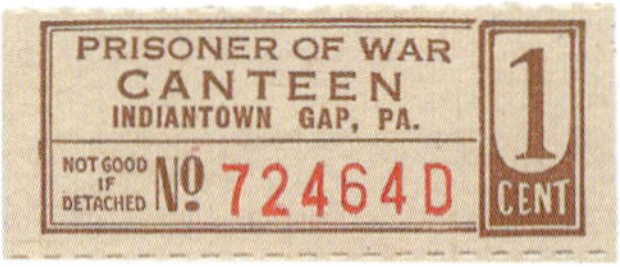

PA-7-2-1

PA-7-2-5

PA-7-2-10

No.	Denomination	Paper	Ink	Serial	Block	Circ.	Unc.
	Canteen in Blue Bar						
PA-7-1-5	5 cents	white	blue	red	A	150	250
	Canteen No Bar						
PA-7-2-1	1 cent	off white	brown	red	D	150	250
PA-7-2-5	5 cents	pale yellow	red	red	B	150	250
PA-7-2-10	10 cents	orange	red	red	B	150	250

Olmsted Field

This POW camp was located at Olmsted Field in Harrisburg, PA. The POW Camp opened in January 1945 and closed in February 1946, with a total of 281 POWS. The former site is now occupied by Harrisburg Airport and The Pennsylvania State University - The Capital College.

PA-18-1-1a

PA-18-1-1b

PA-18-1-5a

PA-18-1-5b

PA-18-1-10a

PA-18-1-10b

PA-18-1-25b

PA-18-1-50b

No.	Denomination	Paper	Ink	Serial	Block	Circ.	Unc.
PA-18-1-1a	1 cent	orange	blue	red	E	25	40
PA-18-1-1b	1 cent	pale orange	blue	red	F	25	40
PA-18-1-5a	5 cents	orange	blue	red	E	35	55
PA-18-1-5b	5 cents	buff	blue	red	F	35	55
PA-18-1-10a	10 cents	orange	blue	red	E	55	85

PA-18-1-10b	10 cents	pink	blue	red	F	55	85
PA-18-1-25a	25 cents	orange	blue	red	E	90	150
PA-18-1-25b	25 cents	orange	blue	red	F	90	125
PA-18-1-50b	50 cents	pale green	blue	red	F	125	175

Camp Reynolds

Camp Reynolds was formerly Camp Shenango in Mercer County, PA. The name was changed on September 21, 1943. There were 1,800 POWS housed there from April 1945 until January 1946.

PA-3-1-1

PA-3-1-2

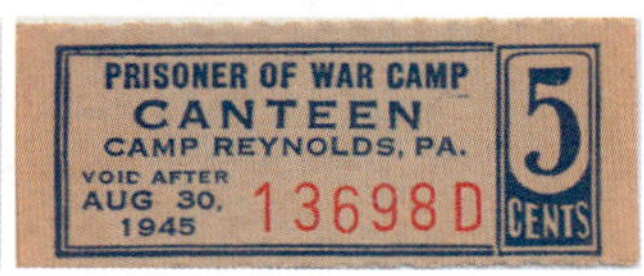

PA-3-1-5

PA-3-1-10

PA-3-1-25

PA-3-2-2

No.	Denomination	Paper	Ink	Serial	Block	Circ.	Unc.
	Aug 30, 1945						
PA-3-1-1	1 cent	buff	blue	red	D	75	175
PA-3-1-2	2 cents	pale orange	blue	red	D	75	150
PA-3-1-5	5 cents	tan	blue	red	D	50	100
PA-3-1-10	10 cents	pale buff	blue	red	D	55	125
PA-3-1-25	25 cents	pale tan	blue	red	D	150	250

	Oct 31, 1945						
PA-3-2-2	2 cents	purple	blue	red	B	150	250

Tobyhanna Military Reservation

In February 1945, the POW compound, located near the present commissary on this reservation, received approximately 300 German POWS. They were used to harvest ice from ponds in the area and worked on local farms in the summer and fall of 1945. A report dated October 1945 stated that 56 POWS were assigned to the local farms, and a logging operation had 41 POWS working there. The ice harvesting had 10 POWS working and others were assigned to work on post in the vast glider storage area. The POWS were repatriated in November 1945, before the onset of winter.

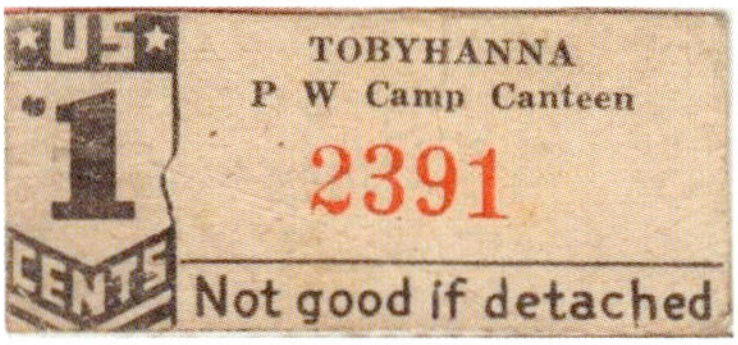

PA-24-1-1

No.	Denomination	Paper	Ink	Serial	Block	Circ.	Unc.
PA-24-1-1	1 cent	tan	black	red		200	400

Valley Forge General Hospital

During World War II there were 244 POWS assigned to this facility in a POW Camp from February 1945 to August 1946. This was the largest general hospital during the war and was located in Phoenixville, PA. The former site in now occupied by Valley Forge Christian College.

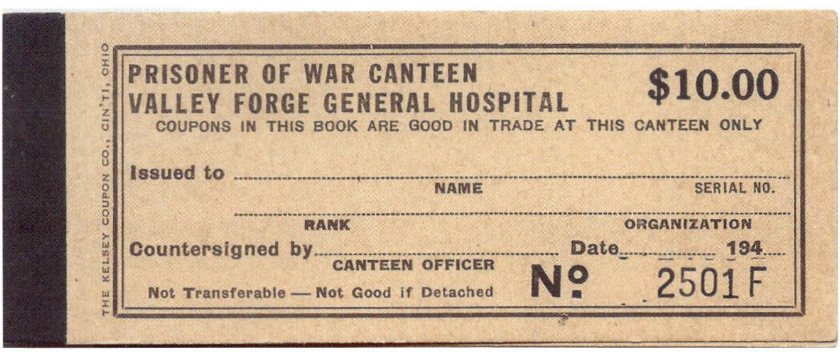

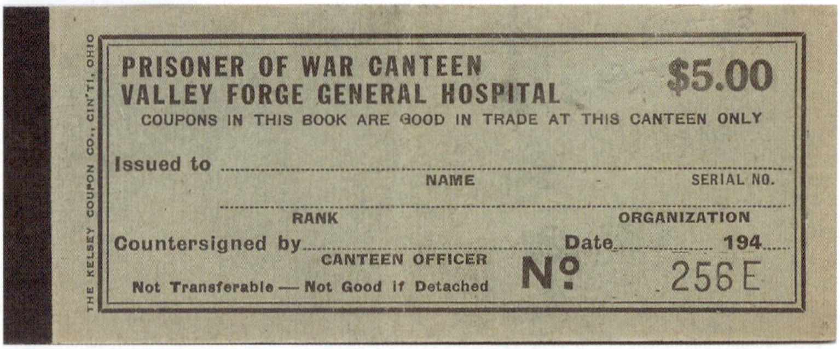

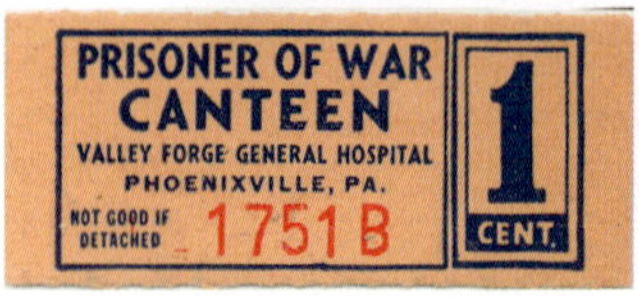

PA-25-1-1a

PA-25-1-1b

PA-25-1-5

PA-25-1-10

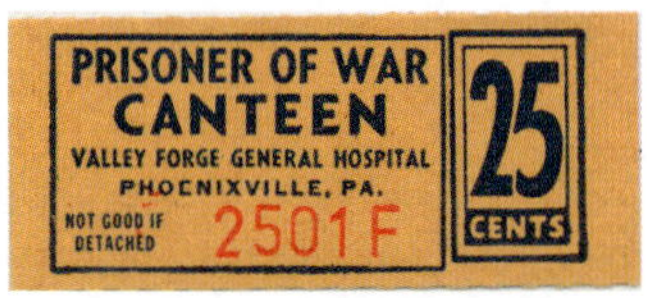

PA-25-1-25

PA-25-1-50

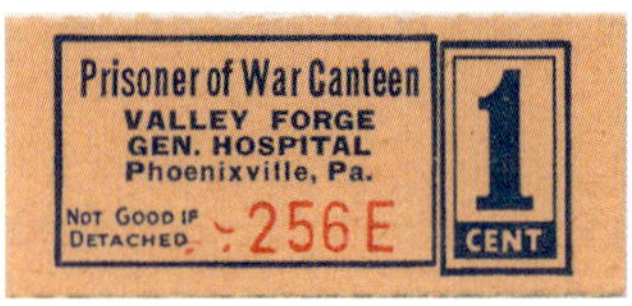

PA-25-2-1

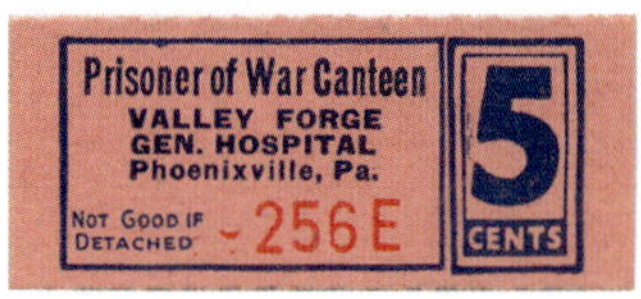

PA-25-2-5

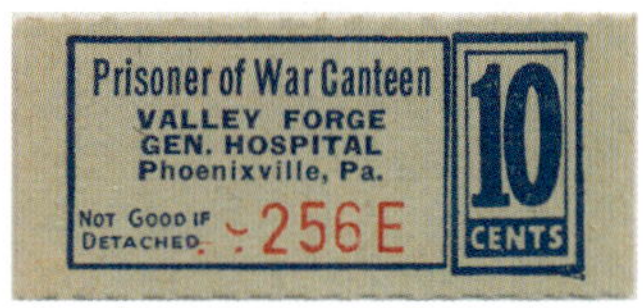

PA-25-2-10

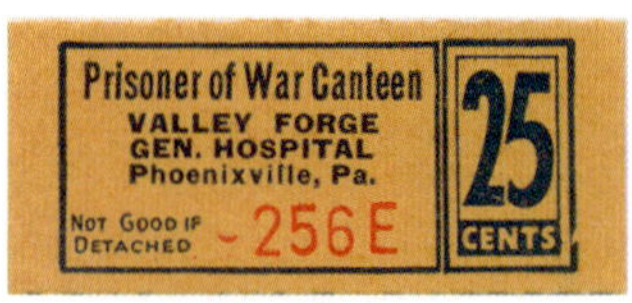

PA-25-2-25

No.	Denomination	Paper	Ink	Serial	Block	Circ.	Unc.
	Canteen on 2nd Line						
PA-25-1-1a	1 cent	pale orange	blue	red	B	25	40
PA-25-1-1b	1 cent error E in Phoenixville like C	pale orange	blue	red	B	100	200
PA-25-1-5	5 cents	rose	blue	red	B	35	55
PA-25-1-10	10 cents	pale green	blue	red	B	55	85

PA-25-1-25	25 cents	tan	blue	red	E	90	150
PA-25-1-50	50 cents	white	blue	red	F	125	175
	Canteen on 1st Line						
PA-25-2-1	1 cent	pale orange	blue	red	E	25	40
PA-25-2-5	5 cents	rose	blue	red	E	35	55
PA-25-2-10	10 cents	pale green	blue	red	E	55	85
PA-25-2-25	25 cents	tan	blue	red	E	90	150

RHODE ISLAND

There are no known POW chits for the three POW camps that were located in Rhode Island. However, there are chits known with the name Narragansett Bay. The three camps were located in and around Jamestown, which is located at the entrance to Narraganssett Bay. The dates published for the camps' openings and closings along with the numbers from Donn are suspect when looking at contemporary newspaper accounts.

	Camp
1	Fort Getty
2	Fort Kearney
3	Fort Wetherill
4	*Harbor Defense of Narragansett Bay

Ft. Getty

Fort Getty was located in Jamestown, RI. In 1941, construction built up the fort to some 40 temporary structures, including fourteen 63-man barracks, five mess halls, and all the support structures required to support the troops. The POW camp opened in July of 1945 and closed in February 1946. During the war, some German POWs were kept for training as post war German civil servants.

Ft. Kearney

This was located across the West passage in Saunderstown. The POW camp opened in March 1944 and closed in March 1946. This fort was also known as "The Factory."

This camp was very unusual in that there were no towers or armed guards. The POWS housed in this camp were handpicked for intelligence and hostility toward the Nazis. Virtually all were conscripts in the army against their will. The 85 POWS were former editors, writers, professors and language specialists; and the captors at the camp were civilian educators and college professors. They were tasked with producing materials in all forms of media to re-educate their countrymen, who were in POW Camps across America. This camp was so successful that at Fort Getty and Fort Wetherill new schools were started to train administrative personal and policemen.

Ft. Wetherill

This camp was located in Jamestown, RI. It was re-activated in September 1940 as part of the Narragansett Bay Harbor defenses. The POW camp opened in August 1945 and closed in November 1945.

There is a contemporary 10 minute video available http://www.youtube.com/watch?v=BNE_E4qPt5M that shows a re-education class.

The last re-education class graduated in December 1945. A total of 1,166 German POWS completed the 60 day courses in schools at Ft. Getty and Ft. Wetherill.

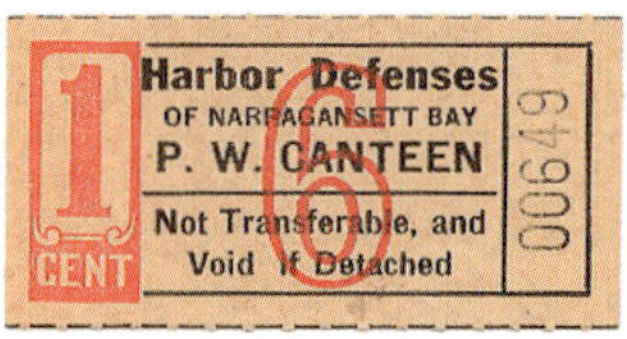

RI-4-1-1

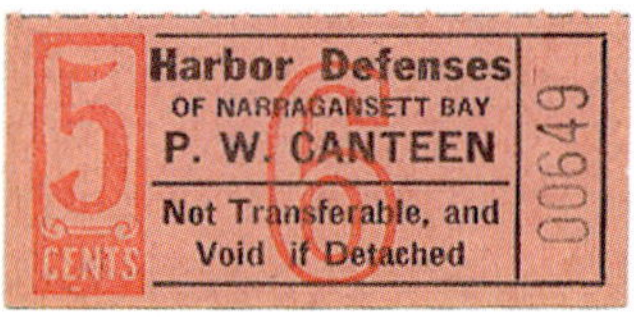

RI-4-1-5

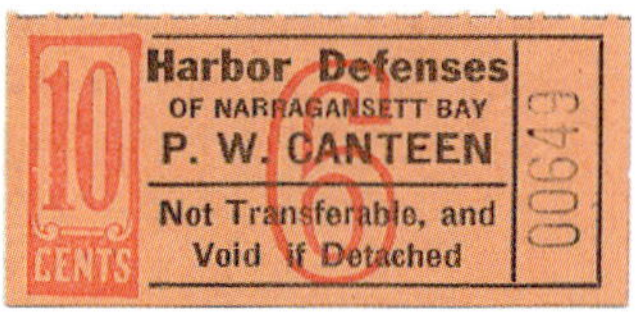

RI-4-1-10

No.	Denomination	Paper	Ink	Serial	Block	Circ.	Unc.
RI-4-1-1	1 cent	tan	black	black	6	75	150
RI-4-1-5	5 cents	rose	black	black	6	100	175
RI-4-1-10	10 cents	pale orange	black	black	6	125	250

SOUTH CAROLINA

	Camp	Base Camp
1	Aiken	Jackson, SC
2	Barnwell	Jackson, SC
3	Bennettsville	Jackson, SC
4	Camden	Jackson, SC
5	Charleston	Jackson, SC
6	Charleston AAB	Jackson, SC
7	Charleston POE	Jackson, SC
8	Columbia AAB	Jackson, SC
9	Coroanco AAB	Jackson, SC
10	Croft	Jackson, SC
11	Florence AAF	Jackson, SC
12	*Fort Jackson	
13	Hampton	Jackson, SC
14	Holly Hill	Jackson, SC
15	Myrtle Beach AAF	Jackson, SC
16	Norway	Jackson, SC
17	Shaw Field	Jackson, SC
18	Stark General Hospital	Jackson, SC
19	Walterboro AAF	Jackson, SC
20	Whitmire	Jackson, SC
21	Witherbee	Jackson, SC

Camp Croft

Camp Croft was located in Spartanburg County near Spartanburg, SC. It was opened for POWS in April 1944, and the last POW left in February 1946. A total of 1,100 POWS were held there.

Ft. Jackson

This camp was located at Columbia, Richland County on a portion of the large training camp for WWII replacement infantrymen. It opened for POWs in June of 1944 and closed in March of 1946.

SC-12-1-1a

SC-12-1-1b

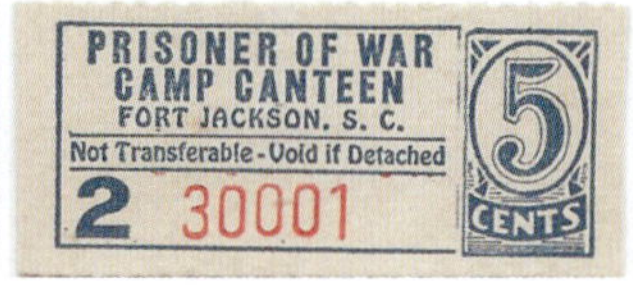

SC-12-1-5

SC-12-1-10

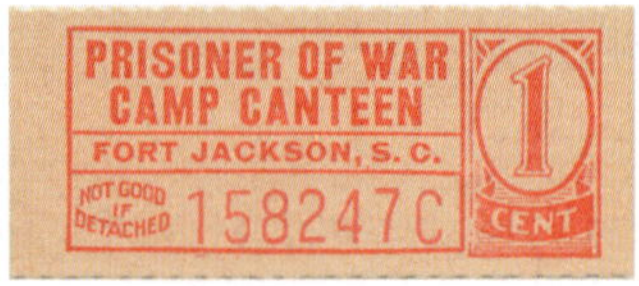

SC-12-2-1

SC-12-2-5

SC-12-2-10

SC-12-2-25

No.	Denomination	Paper	Ink	Serial	Block	Circ.	Unc.
SC-12-1-1a	1 cent	pale green	blue	red	2	25	40
SC-12-1-1b	1 cent	blue	blue	red	3	25	40
SC-12-1-5	5 cents	buff	blue	red	2	35	55
SC-12-1-10	10 cents	orange	blue	red	2	55	85
	Ft. Jackson Between Lines						
SC-12-2-1	1 cent	pale orange	red	red	C	25	40
SC-12-2-5	5 cents	pale orange	red	red	C	35	55
SC-12-2-10	10 cents	mauve	red	red	C	55	85
SC-12-2-25	25 cents	orange	red	red	C	90	125

SOUTH DAKOTA

There were only sub-camps located in South Dakota. Any chits that were used were probably from their base camps located in other states.

	Camp	Base Camp
1	Belle Fourche	Scottsbluff, NE
2	Ft. Meade	Robinson, NE
3	Ft. Meade II	Robinson, NE
4	Provo	Scottsbluff, NE

5	Sioux Falls	Algona, IA
6	Yankton	Algona, IA

TENNESSEE

	Camp	Base Camp
1	*Camp Forrest	
2	*Crossville	
3	Jackson	Forrest, TN
4	Lawrenceburg	Forrest, TN
5	*Memphis ASF	
6	McKellar Field	Forrest, TN
7	Spencer Range	Forrest, TN
8	Tellico Plains	Forrest, TN
9	Thayer General Hospital	Forrest, TN
10	*Tyson	

Camp Crossville

Camp Crossville opened for its first POWs in November 1942. Over 1,500 POWs were held here up to when the camp closed in December 1945. The POWs worked on local farms as well as at the Tennessee Valley Authority or the University of Tennessee.

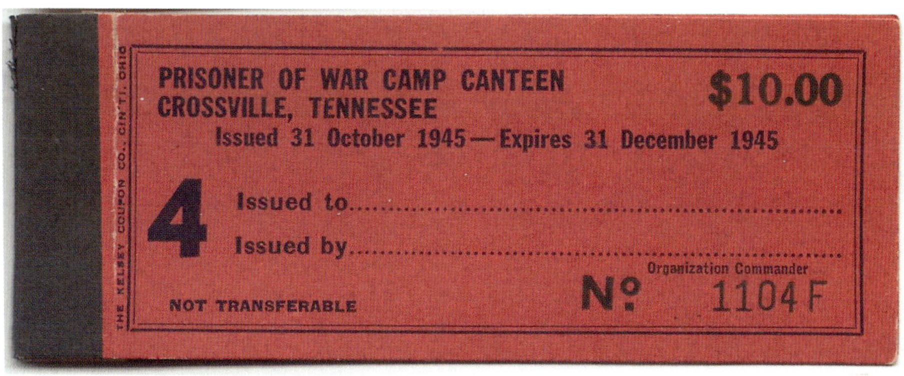

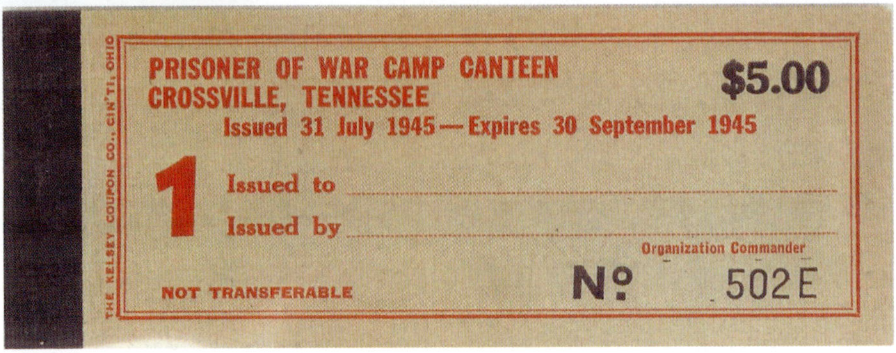

TN-2-1-1

TN-2-1-5

TN-2-1-10

TN-2-2-3

TN-2-3-1

TN-2-3-3

TN-2-3-5

TN-2-3-10

TN-2-3-25

TN-2-4-1

TN-2-4-3

TN-2-4-5

TN-2-4-10

TN-2-4-25

TN-2-5-1

TN-2-5-3

TN-2-5-5

TN-2-5-10

TN-2-5-25

TN-2-5-50

No.	Denomination	Paper	Ink	Serial	Block	Circ.	Unc.
	Exchange						
TN-2-1-1	1 cent	buff	red	red	B	25	40
TN-2-1-5	5 cents	buff	red	red	B	35	55
TN-2-1-10	10 cents	buff	red	red	B	55	85
	May						
TN-2-2-3	3 cents	yellow	red	red	A	75	150
	September						
TN-2-3-1	1 cent	white	red	red	A	25	40
TN-2-3-3	3 cents	orange	red	red	A	75	150
TN-2-3-5	5 cents	pink	red	red	E	35	55
TN-2-3-10	10 cents	gray	red	red	E	55	85
TN-2-3-25	25 cents	buff	red	red	E	90	125
	November						
TN-2-4-1	1 cent	gray	blue	red	E	25	40
TN-2-4-3	3 cents	orange	blue	red	E	115	150
TN-2-4-5	5 cents	white	blue	red	E	35	55
TN-2-4-10	10 cents	red	blue	red	E	55	85
TN-2-4-25	25 cents	pink	blue	red	E	90	125
	December						
TN-2-5-1	1 cent	pale green	blue	red	F	25	40
TN-2-5-3	3 cents	buff	blue	red	F	115	150

TN-2-5-5	5 cents	red	blue	red	F	35	55
TN-2-5-10	10 cents	white	blue	red	F	55	85
TN-2-5-25	25 cents	pink	blue	red	F	90	125
TN-2-5-50	50 cents	gray	blue	red	F	115	150

Camp Forrest

Camp Forrest had a dual purpose, both as an internment camp and a POW camp. It was the first internment camp in our nation. Beginning in January 1942, over 800 alien civilians were housed there. These civilian detainees were arrested after the outbreak of WWII, under a program called "Alien Enemy Control." Many of these internees were incarcerated without legal process.

The POW camp at Camp Forrest opened May 15, 1942. In order to make more room for POWs in the camp, which eventually held 24,000 prisoners, the internment camp was closed in early 1943, with the internees being relocated to other internment camps in the United States. The POW camp eventually closed in May 1946.

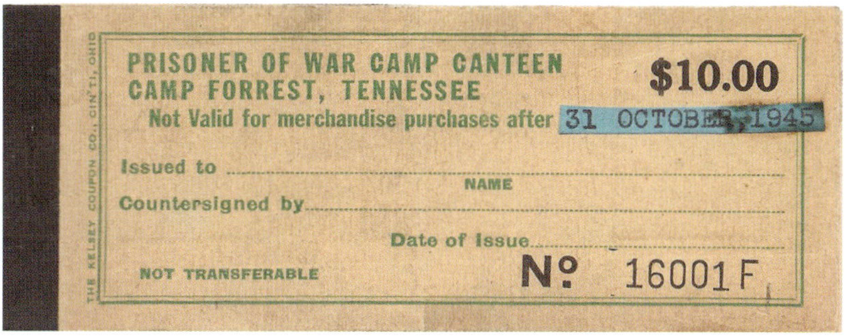

TN-1-1-1

TN-1-1-5

TN-1-1-10

TN-1-1-25

TN-1-2-1a

TN-1-2-1b

TN-1-2-5

TN-1-2-10

TN-1-2-25

TN-1-3-1

TN-1-4-1

TN-1-5-1a

TN-1-5-1b

TN-1-5-5

TN-1-5-10

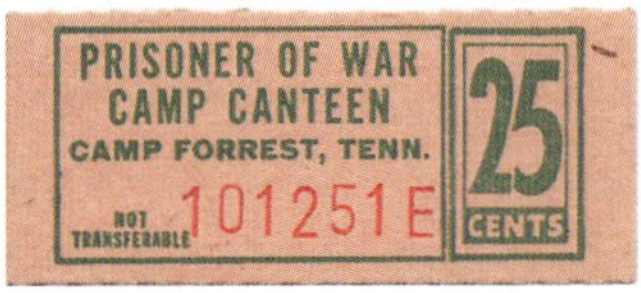

TN-1-5-25

TN-1-5-50

No.	Denomination	Paper	Ink	Serial	Block	Circ.	Unc.
	Camp Exchange						
TN-1-1-1	1 cent	orange	blue	red	C	25	40
TN-1-1-5	5 cents	pink	blue	red	C	35	55
TN-1-1-10	10 cents	beige	blue	red	C	55	85
TN-1-1-25	25 cents	green	blue	red	C	90	125
	25 June 1945 **Denomination in square**						
TN-1-2-1a	1 cent	gray	red	red	D	25	40
TN-1-2-1b	1 cent	orange	red	red	D	25	40
TN-1-2-5	5 cents	buff	red	red	D	35	55
TN-1-2-10	10 cents	beige	red	red	D	55	85
TN-1-2-25	25 cents	pink	red	red	D	90	125
	25 June 1945 **Denomination in circle**						
TN-1-3-1	1 cent	orange	red	red	D	90	125
	25 Dec 1945						
TN-1-4-1	1 cent	pale orange	green	red	E	90	125
	Camp Canteen						

TN-1-5-1a	1 cent	pale orange	green	red	E	25	40
TN-1-5-1b	1 cent	gray	green	red	F	25	40
TN-1-5-5	5 cents	white	green	red	F	35	55
TN-1-5-10	10 cents	buff	green	red	F	55	85
TN-1-5-25	25 cents	pale orange	green	red	F	90	125
TN-1-5-50	50 cents	green	green	red	F	115	150

Memphis ASF

When the prisoner of war camp was established early in 1944 at the depot in Memphis, it was a branch of the Como, Mississippi camp and housed about 250 Italian POWs. Around the first of May 1944, approximately 300 German POWs replaced them. By September 1945, there were 1,912 prisoners. They were involved with manual labor, mowing and grounds maintenance, and cafeteria work. Some eight hundred POWs worked in the warehouses, moving a great deal of the tonnage handled by the depot. The POW camp was closed in May of 1946.

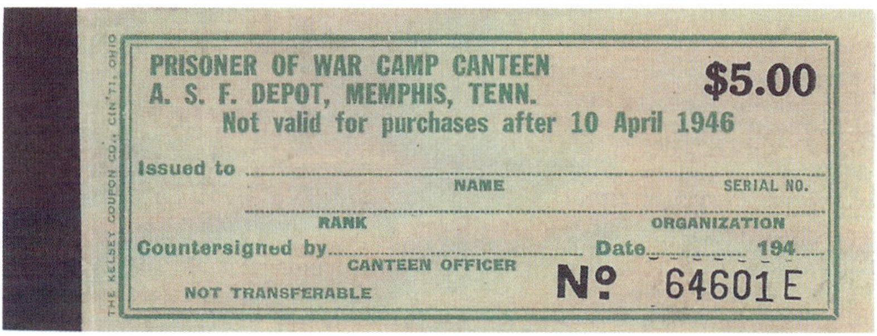

TN-5-1-1a

TN-5-1-1b

TN-5-1-1c

TN-5-1-1d

TN-5-1-5a

TN-5-1-5b

TN-5-1-1c

TN-5-1-5d

TN-5-1-10a

TN-5-1-10c

TN-5-1-25a

TN-5-2-1

TN-5-2-5

TN-5-2-10

TN-5-3-1

TN-5-3-5

TN-5-3-10

No.	Denomination	Paper	Ink	Serial	Block	Circ.	Unc.
TN-5-1-1a	1 cent	gray	blue	red	A	25	40
TN-5-1-1b	1 cent	peach	blue	red	A	25	40
TN-5-1-1c	1 cent	peach	red	red	E	25	40
TN-5-1-1d	1 cent	green	green	red	E	25	40
TN-5-1-5a	5 cents	peach	blue	red	A	35	55
TN-5-1-5b	5 cents	orange	blue	red	A	35	55
TN-5-1-5c	5 cents	cream	red	red	E	35	55
TN-5-1-5d	5 cents	pink	green	red	E	35	55
TN-5-1-10a	10 cents	green	blue	red	A	55	85
TN-5-1-10b	10 cents	pink	blue	red	A	55	85

TN-5-1-10c	10 cents	gray	red	red	E	55	85
TN-5-1-10d	10 cents	pale orange	green	red	E	55	85
TN-5-1-25a	25 cents	cream	red	red	E	90	125
TN-5-1-25b	25 cents	buff	green	red	E	90	125
	June 25, 1945						
TN-5-2-1	1 cent	white	blue	red	A	25	40
TN-5-2-5	5 cents	orange	blue	red	A	35	55
TN-5-2-10	10 cents	yellow	blue	red	A	55	85
	December 25, 1945						
TN-5-3-1	1 cent	green	blue	red	A	25	40
TN-5-3-5	5 cent	gray	blue	red	A	35	55
TN-5-3-10	10 cents	tan	blue	red	A	55	85

TEXAS

	Camp	Base Camp
1	Alto	Fannin, TX
2	Alvin	Swift, TX
3	Amarillo AAF	Hereford, TX
4	Anahuac	Huntsville, TX
5	Angleton	Swift, TX
6	Ashburn General Hospital	Howze, TX
7	Atlanta	
8	Bannister	Fannin, TX
9	Bay City	Swift, TX
10	Big Springs AAF	Bowie, TX
11	Biggs Field	Ft. Bliss, TX
12	Baumont General Hospital	
13	*Brady	
14	Bryan	Hearne, TX
15	*Camp Barkeley	
16	*Camp Bowie	
17	*Camp Fannin	

18	*Camp Hood	
19	*Camp Hulen	
20	*Camp Howze	
21	*Camp Maxey	
22	*Camp Swift	
23	*Camp Wallace	
24	*Camp Wolters	
25	Canutillo	Ft. Bliss, TX
26	Center	Fannin, TX
27	Chance Plantation	Hearne, TX
28	Childress AAF	Hereford, TX
29	China	Huntsville, TX
30	Chireno	Fannin, TX
31	Cleburne	Mexia, TX
32	Corpus Christi	Ft. Sam Houston,
33	Corsicana	Mexia, TX
34	Denison	Howze, TX
35	Dumas	Hereford, TX
36	Eagle Lake	Swift, TX
37	El Champo	Swift, TX
38	Fabens	Bliss, TX
39	Forney	Mexia, TX
40	*Fort Bliss	
41	*Fort Clark	
42	*Fort Crockett	
43	*Fort Sam Houston	
44	Fort Worth QMD	Bowie, TX
45	Ganado	Swift, TX
46	Garwood	Swift, TX
47	Harmon General Hospital	Fannin, TX
48	*Hearne	
49	*Hereford	

50	Humble	
51	*Huntsville	
52	Kaufman	Media, TX
53	Kenedy	Ft. Sam Houston,
54	Kirbyville	Huntsville, TX
55	Liberty	Huntsville, TX
56	Lubbock AAF	Hereford, TX
57	Lufkin	Fannin, TX
58	McClosky General Hospital	Hood, TX
59	*McLean	
60	*Mexia	
61	Milam	Fannin, TX
62	Mont Belvieu	Huntsville, TX
63	Navasota	Huntsville, TX
64	*North Camp Hood	
65	Orange	Huntsville, TX
66	Patroon	Fannin, TX
67	Princeton	Howze, TX
68	Pyote AAF	Rosewell, NM
69	Rosenberg	Swift, TX
70	Ft. D.A. Russell	
71	San Augustine	Fannin, TX
72	Wharton	Swift, TX
73	White Rock Lake	Mexia, TX
74	Ysleta	Bliss, TX

Camp Barkeley POW Base Camp

Camp Barkeley was eleven miles southwest of Abilene. It was originally planned as a temporary camp but, with the start of World War II, it became one of Texas' largest military installations. The camp was named for David B. Barkley, a native Texan who was awarded the Medal of Honor in World War I. A typing error apparently caused the spelling change. Construction started in December 1940. The government leased 70,229 acres, with an option to buy additional land, to accommodate the camp. The cost of the camp was originally estimated at $4 million, but when completed in July

1941, the cost was $7 million. The camp also served as a camp for German prisoners of war. On February 1, 1944, the 1846th Unit POW Camp was activated under the command of Lt. Col. Harry Slaughter. At its peak in March 1945, the POW camp housed 840 prisoners. Two months after the camp opened, twelve prisoners escaped, but all were recaptured within a week. On April 30, 1945, Camp Barkeley was deactivated.

TX-15-1-5

TX-15-1-10

No.	Denomination	Paper	Ink	Serial	Block	Circ.	Unc.
TX-15-1-5	5 cents	red	black	black	B	75	150
TX-15-1-10	10 cents	green	black	red	E	50	100

Camp Bowie POW Base Camp

The Camp Bowie prisoner of war camp located in Brown County was activated on July 10, 1943. The camp was built in 30 days. It was located in the valley immediately below the dam on the Brownwood Country Club property. The first prisoners arrived from Europe in August 1943. After 30 days, the original population of 70 or so had climbed to 1,800. By September 1943 there were 2,724 prisoners. Most of them were members the Afrika Korps. The prisoners came by train and were unloaded at the

Brownwood Sante Fe Depot, and they marched to the campsite, a distance of 3½ miles. The prisoners were dressed in field gear, including knee-high tapped boots. To the cadence of the German Army marching songs, they sang their way to the compound. In 1945, the camp was changed from one for army POWs to a camp for German naval personnel.

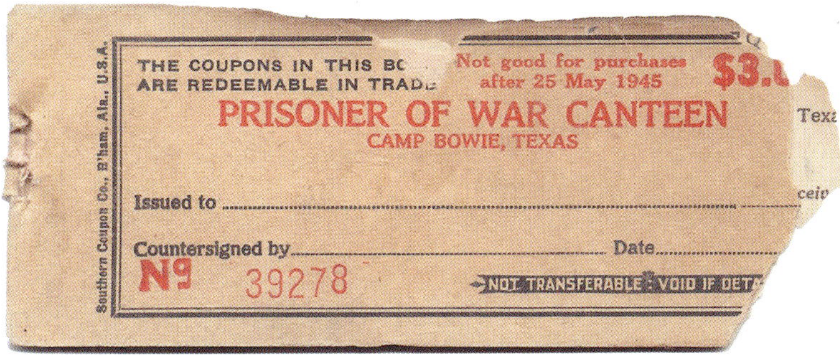

TX-16-1-1a

TX-16-1-1b

TX-16-1-5a

TX-16-1-5b

TX-16-1-5c

TX-16-1-10a

TX-16-1-10b

TX-16-1-10c

TX-16-2-1

TX-16-2-5

No.	Denomination	Paper	Ink	Serial	Block	Circ.	Unc.
TX-16-1-1a	1 cent	cream	blue	red		40	80
TX-16-1-1b	1 cent	cream	red	red		40	80
TX-16-1-5a	5 cents	green	blue	red		60	120
TX-16-1-5b	5 cents	pink	blue	red		60	120
TX-16-1-5c	5 cents	blue	red	red		60	120
TX-16-1-10a	10 cents	orange	black	red		80	160
TX-16-1-10b	10 cents	blue	blue	red		80	160
TX-16-1-10c	10 cents	gray	red	red		80	160
	Canteen in Bar						
TX-16-2-1	1 cent	cream	blue	red	B	25	40
TX-16-2-5	5 cents	green	blue	red	B	40	55

Camp Brady POW Base Camp

The camp was composed of 1,200 German SS and Gestapo soldiers. It was 360 acres, which included 20 buildings and was built at a cost of $2 million. It was activated February 1944 and closed June 1945. The largest building was the 15 bed hospital. The camp capacity was 3,000. It remained active throughout the latter half of World War II.

TX-13-1-1

TX-13-2-1

TX-13-2-2

TX-13-3-1

TX-13-3-2

No.	Denomination	Paper	Ink	Serial	Block	Circ.	Unc.
	In Merchandise						
TX-13-1-1	1 cent expedient	peach	black	red		100	200
	With POW Overprint						
TX-13-2-1	1 cent expedient	peach	black	red		200	400
TX-13-2-2	2 cents expedient	mauve	black	red		200	400
	Camp Canteen						
TX-13-3-1	1 cent	yellow	black	red	A	150	250
TX-13-3-2	2 cents	orange	black	red	A	150	250

Camp Fannin POW Base Camp

The camp, near Tyler in Smith County, opened for POWs in September 1943 and closed in June 1946. The area where Camp Fannin existed was returned to non-military use during 1946. A section of the land was handed over to the state of Texas, where the once military hospital was transformed to the East Texas Tuberculosis Sanatorium, later the University of Texas Health Center at Tyler. Other buildings were sold to commercial and development companies. Many buildings still stand today, most of which are used as locations for businesses and warehouses.

TX-17-1-1

TX-17-1-5

TX-17-1-10

TX-17-1-25

No.	Denomination	Paper	Ink	Serial	Block	Circ.	Unc.
TX-17-1-1	1 cent	pink	black	red	A	150	250
TX-17-1-5	5 cents	cream	black	red	A	150	250
TX-17-1-10	10 cent	peach	black	red	A	150	250
TX-17-1-25	25 cents	orange	black	red	A	200	300

Camp Hood POW Base Camp

Camp Hood, located in Bell County, opened for POWs in June 1943 and closed three years later in June 1946. A total of 1,641 German POWs were housed at the camp.

TX-18-1-1

No.	Denomination	Paper	Ink	Serial	Block	Circ.	Unc.
TX-18-1-1	1 cent	pink	blue	red	E	150	250

Camp Hulen POW Base Camp

The camp opened for POWs in February 1944 and closed in September 1945. A total of only 190 German POWs were housed in this camp.

TX-19-1-1

TX-19-1-5

No.	Denomination	Paper	Ink	Serial	Block	Circ.	Unc.
TX-19-1-1	1 cent	pink	red	black	E	150	250
TX-19-1-5	5 cents	gray	red	black	E	150	250

Camp Maxey POW Base Camp

Camp Maxey, near Paris, opened in June of 1943 and held up to 5,237 POWs. The camp closed in June of 1946.

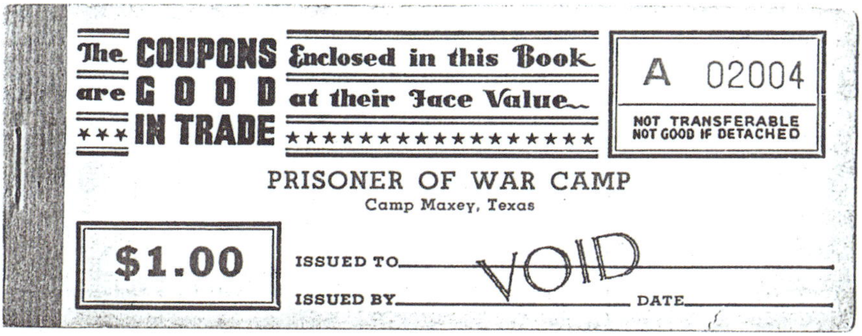

TX-21-1-1

TX-21-1-5

TX-21-1-10

TX-21-2-1a

TX-21-2-1b

TX-21-2-5

TX-21-2-10

TX-21-2-25

No.	Denomination	Paper	Ink	Serial	Block	Circ.	Unc.
	Trade Coupon						
TX-21-1-1	1 cent	pink	black & red	red	A	50	100
TX-21-1-5	5 cents	peach	black & red	red	A	75	150
TX-21-1-10	10 cents	green	black & red	red	A	100	200
	Camp Canteen						
TX-21-2-1a	1 cent	orange	black	black	A	75	150
TX-21-2-1b	1 cent	rose	red	black	E	75	150
TX-21-2-5	5 cents	green	red	black	E	100	200

TX-21-2-10	10 cents	gray	red	black	E	100	200
TX-21-2-25	25 cents	orange	red	black	E	150	250

Camp Swift POW Base Camp

Camp Swift, located in Bastrop County, opened for POWs in June of 1943. The peak population was 2,205, and the last POWs left in February 1946. The Camp housed mostly German soldiers captured in North Africa. There were also 300 Russians confined, who were conscripted to fight on the German side, but nonetheless had to be segregated from the Germans.

TX-22-1-1

TX-22-2-1

TX-22-2-5

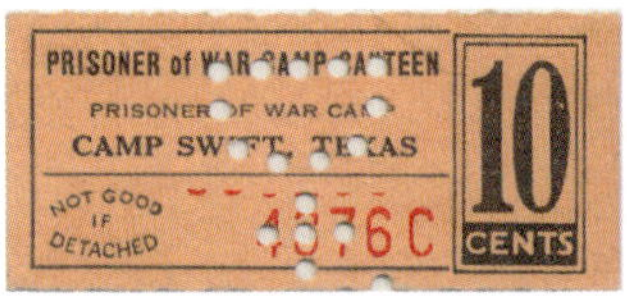

TX-22-2-10

TX-22-2-25

No.	Denomination	Paper	Ink	Serial	Block	Circ.	Unc.
	Camp Exchange						
TX-22-1-1	1 cent	tan	black	red		100	200
	Camp Canteen						
TX-22-2-1	1 cent	mauve	black	red	C,E	100	200
TX-22-2-5	5 cents	yellow	black	red	C	125	225
TX-22-2-10	10 cents	peach	black	red	C	150	250
TX-22-2-25	25 cents	red	black	red	C	200	400

Camp Wolters POW Base Camp

The POW camp at “Old Camp Wolters” opened in March 1944. It was located in Palo Pinto County near Mineral Springs. The original buildings of “Old Camp Wolters” were converted into a POW camp for German prisoners from North Africa. The prison camp was closed on August 15, 1946, as the last prisoners were returned to their homeland.

TX-24-1-1

TX-24-1-5

TX-24-2-1

TX-24-2-10

TX-24-2-25

TX-24-2-50

No.	Denomination	Paper	Ink	Serial	Block	Circ.	Unc.
	PW Camp						
TX-24-1-1	1 cent	orange	black	red		100	200
TX-24-1-5	5 cents	gray	black	red		125	250
	Prisoner of War						
TX-24-2-1	1 cent	beige	black	red		100	200
TX-24-2-5	5 cents						
TX-24-2-10	10 cents	pink	black	red		150	300
TX-24-2-25	25 cents	beige	black	red		175	350
TX-24-2-50	50 cents	pink	black	red		200	400

Fort Bliss POW Base Camp

Fort Bliss in El Paso County was originally a tent camp, with room for 200 prisoners. In 1942, it was enlarged to accommodate 1,300. Camp Bliss was home for German and Italian prisoners for approximately 18 months and housed an average of 3,000 prisoners each month it was open. It closed on June 15, 1946, when the last prisoners were sent back to Europe.

TX-40-1-1

No.	Denomination	Paper	Ink	Serial	Block	Circ.	Unc.
TX-40-1-1	1 cent trade coupon	pink	black & red	red		200	400

Ft. Clark POW Base Camp

Fort Clark, Kinney County was near Brackettville, TX. The German POW sub-camp on the 4,000 acre reservation for 199 POWs opened in July 1945 and closed in September 1945.

TX-41-1-1

TX-41-1-5

No.	Denomination	Paper	Ink	Serial	Block	Circ.	Unc.
TX-41-1-1	1 cent	white	blue	red	A	100	200
TX-41-1-5	5 cents	pink	blue	red	C	100	200

Ft. Crockett POW Base Camp

The fort was located in Galveston, TX. It opened in 1943 and closed in May of 1946. Fort Crockett became a prisoner of war camp in 1943 when the first group of 165 POWs arrived in Galveston. The compound was from 53rd Street to 57th Street and from Avenue Q to Seawall Boulevard. The compound fence went across the Boulevard, down to the beach and across the beach into the water. A total of 650 POWs were detained at the camp until it was deactivated in 1946.

TX-42-1-5

TX-42-1-10

TX-42-1-25

TX-42-1-50

TX-42-2-5

TX-42-2-10

TX-42-2-25a

TX-42-2-25b

TX-42-2-50

TX-42-3-50

No.	Denomination	Paper	Ink	Serial	Block	Circ.	Unc.
	Trade Coupon						
TX-42-1-5	5 cents	beige	black & red	red		100	200
TX-42-1-10	10 cents	gray	black & red	red		125	250
TX-42-1-25	25 cents	pink	black & red	red		150	300
TX-42-1-50	50 cents	tan	black & red	red		200	400
	Canteen 2nd Line						
TX-42-2-5	5 cents	orange	black	black		100	200
TX-42-2-10	10 cents	beige	black	red		125	250
TX-42-2-25a	25 cents	beige	black	black		150	300
TX-42-2-25b	25 cents	beige	black	red		150	300
TX-42-2-50	50 cents	red	black	black		200	400
	Canteen 1st Line						
TX-42-3-50	50 cents	orange	black & red	red		200	400

Ft. Sam Houston POW Base Camp

Located in San Antonio, the first "several hundred German prisoners of war were scheduled to arrive in late October 1943 at Fort Sam," reported the *San Antonio Express* of Oct. 14, 1943. The camp was officially established on Oct. 21, 1943 in the six northernmost rows of tents in Dodd Field. It was surrounded by a double chain-link fence topped with barbed wire and watched over by military police in guard towers around the perimeter. The camp closed in July of 1946. At its peak the camp held 2,979 POWS.

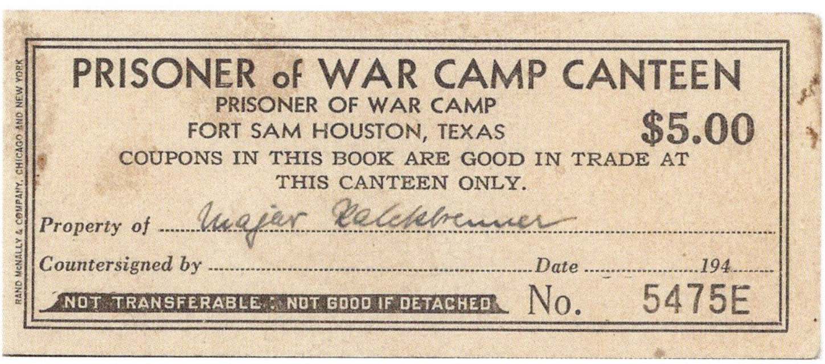

TX-43-1-1

No.	Denomination	Paper	Ink	Serial	Block	Circ.	Unc.
TX-43-1-1	1 cent	green	blue	blue	B	150	300

Hearne POW Base Camp

Opened in June of 1943 in Robertson County, Hearne was chosen to house 4,800 prisoners in its POW base camp due to its flat terrain, distance from the coastline, railroad access, sparse population, and the local need for farm laborers. The camp closed in February 1946. Up to 2,929 POWs were housed at this camp, with the majority being NCOs.

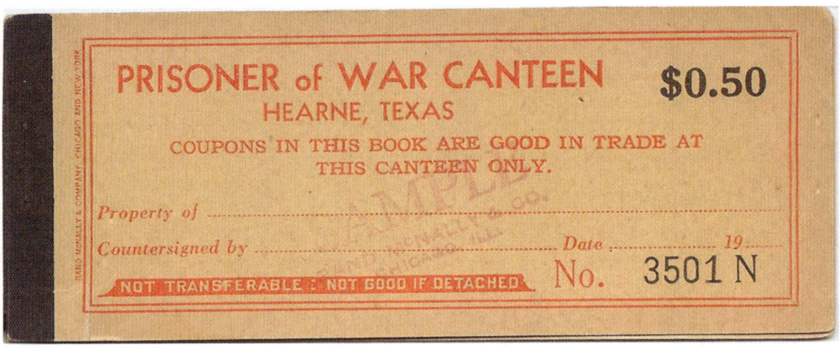

TX-48-1-1a

TX-48-1-1b

TX-48-1-5

TX-48-2-1

No.	Denomination	Paper	Ink	Serial	Block	Circ.	Unc.
	Denomination on Right						
TX-48-1-1a	1 cent	gray	red	black	N	100	150
TX-48-1-1b	1 cent	blue	red	black	N	100	150
TX-48-1-5	5 cents	rose	red	black	N	150	200
	Denomination on Left						
TX-48-2-1	1 cent	gray	black	red		100	150

Hereford POW Base Camp

During World War II, Hereford in Deaf County became the site of a prisoner of war camp named the Hereford Internment Area. Covering a section of land, the camp was constructed in 1942 at a cost of two million dollars. The first American Military Police unit arrived in 1943. Italian captives arrived in April of that same year. By September of 1943, the camp contained 4,000 prisoners. The camp closed in February 1946.

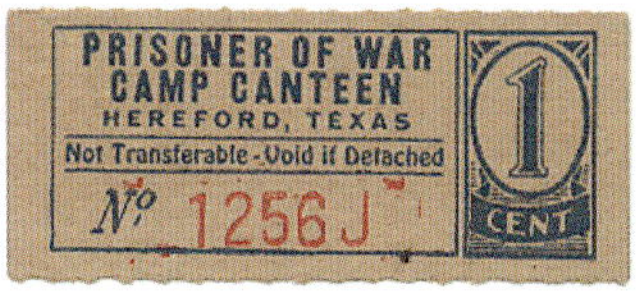

TX-49-1-1

TX-49-2-25

No.	Denomination	Paper	Ink	Serial	Block	Circ.	Unc.
	With No.						
TX-49-1-1	1 cent	tan	blue	red	B,H	100	200
	Without No.						
TX-49-2-25	25 cents	rose	blue	red	C	150	300

Huntsville POW Base Camp

This camp, located eight miles from Huntsville, contained 837 acres taken by eminent domain. Construction began in the summer of 1942, with the camp opening to primarily German POWs in October 1943. This was the first camp of many camps that were built in Texas, and it was one of the first in the country. The American government ran a top secret re-education program for German and Japanese prisoners at the camp. Huntsville was the only camp in the country with a Japanese re-education program. The camp closed in September 1945.

TX-51-1-1

TX-51-1-5

TX-51-1-10

TX-51-1-25

TX-51-2-1

TX-51-2-5

TX-51-2-10

No.	Denomination	Paper	Ink	Serial	Block	Circ.	Unc.
TX-51-1-1	1 cent	tan	blue	red	B	100	200
TX-51-1-5	5 cents	mauve	blue	red	B	125	225
TX-51-1-10	10 cents	green	blue	red	B,C	150	250
TX-51-1-25	25 cents	red	blue	red	B	175	300
	Large Canteen						
TX-51-2-1	1 cent	pink	blue	red	A,C,E	25	40
TX-51-2-5	5 cents	orange	blue	red	A,C,E	35	55
TX-51-2-10	10 cents	green	blue	red	A,C	50	85
TX-51-2-25	25 cents	white	blue	red	C	80	125

McLean POW Base Camp

A prisoner of war camp was established here during World War II. Construction began in September 1942, and American military personnel and German prisoners began to arrive in July 1943. By October 1944, the camp exceeded its official capacity of 3,000 prisoners. The camp was closed in October 1945.

TX-59-1-1

TX-59-1-10

TX-59-2-1

No.	Denomination	Paper	Ink	Serial	Block	Circ.	Unc.
	Denomination on Left						
TX-59-1-1	1 cent	peach	black	red		100	200
TX-59-1-10	10 cents	orange	black	red		150	300
	Denomination on Right						
TX-59-2-1	1 cent	peach	black	red	A	150	300

Mexia POW Base Camp

The camp, located in Limestone County, opened in June of 1943. The residents of Mexia watched 1,850 German POWs, former members of the Africa Korps, disembark from the troop train and march in formation, in cadence, to the camp situated four miles west of town. The incoming prisoners included Frenchmen, who had been pressed into the German Army, along with a platoon of Arabs from the North African campaign. The rest of the contingent consisted of three hundred naval officers, including an admiral and nearly one thousand German Army officers, four generals among them. The camp closed in December 1945. At its peak the camp held 2,029 POWs.

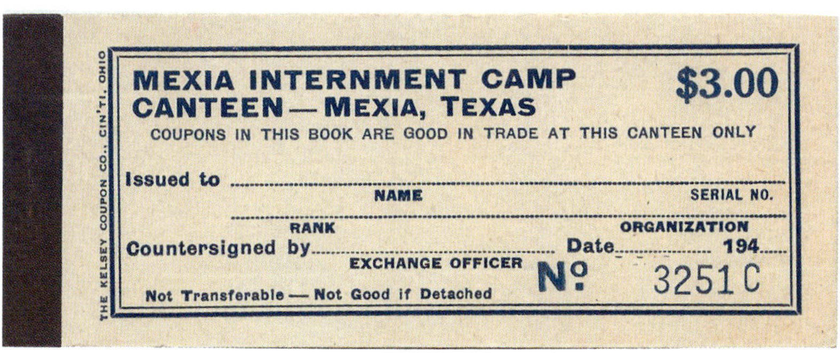

TX-60-1-1a

TX-60-1-1b

TX-60-1-5a

TX-60-1-5b

TX-60-1-10a

TX-60-1-10b

TX-60-1-25a

TX-60-1-25b

TX-60-1-50b

TX-60-2-1

TX-60-3-1

TX-60-3-5

TX-60-3-10

No.	Denomination	Paper	Ink	Serial	Block	Circ.	Unc.
	Prisoner of War						
TX-60-1-1a	1 cent	orange	black	red	A,C,D	25	40
TX-60-1-1b	1 cent	orange	blue	red	D,E	25	40
TX-60-1-5a	5 cents	buff	black	red	A	35	55
TX-60-1-5b	5 cents	buff	blue	red	E	35	55
TX-60-1-10a	10 cents	beige	black	red	A,D	55	85
TX-60-1-10b	10 cents	beige	blue	red	E	55	85
TX-60-1-25a	25 cents	pink	black	red	D	90	125
TX-60-1-25b	25 cents	pink	blue	red	E	90	125
TX-60-1-50b	50 cents	green	blue	red	E	110	150
	P.O.W.						
TX-60-2-1	1 cent	white	blue	red	A	25	40
TX-60-2-5	5 cents	green	blue	red	A	35	55
TX-60-2-10	10 cents	pink	blue	red	A	55	85
	P.O.W. Camp						
TX-60-3-1	1 cent	buff	blue	red	A	25	40
TX-60-3-5	5 cents	pink	blue	red	A	35	55
TX-60-3-10	10 cents	green	blue	red	A	55	85

North Camp Hood POW Base Camp

The original facilities provided housing and training sites for nearly 38,000 troops. In January 1943, 34,943 acres (141 km²) in Coryell County near Gatesville, Texas were purchased. The site near Gatesville was known as the sub-camp and later as North Camp Hood. During the war years, North Camp Hood housed nearly 40,000 troops and 4,000 prisoners of war. The camp closed in June of 1946.

TX-64-1-1

TX-64-1-5

No.	Denomination	Paper	Ink	Serial	Block	Circ.	Unc.
TX-64-1-1	1 cent	pink	red	black	A	150	300
TX-64-1-5	5 cents	green	red	black	A	200	400

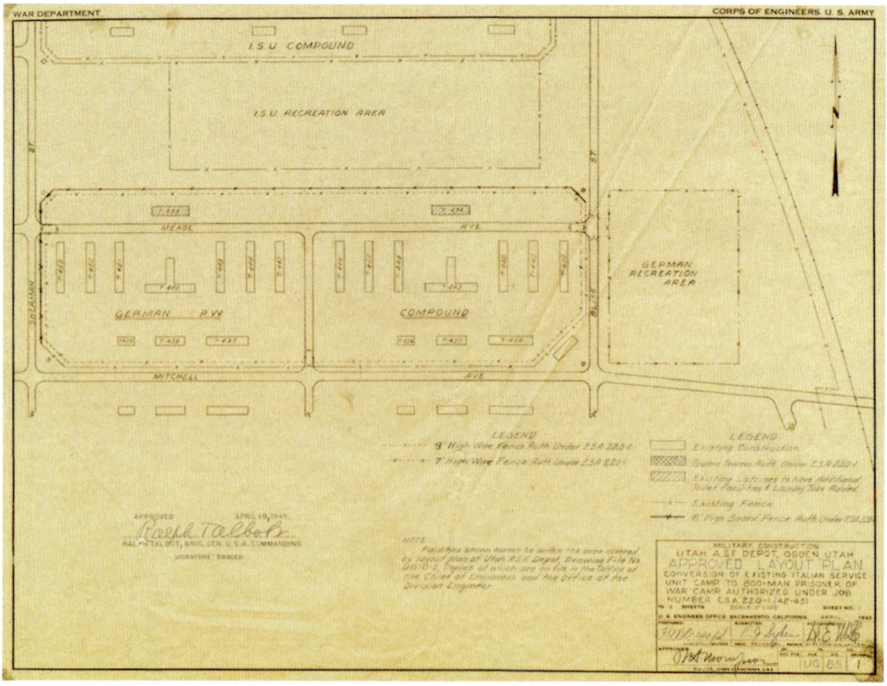

Approved Layout Plan for the conversion of the Utah ASF Depot Italian Service Unit camp to an 800 person German POW camp.

UTAH

	Camp	Base Camp
1	Bushnell General Hospi-	
2	Clearfield	
3	Desert Chemical Warfare	
4	Ft. Douglas	
5	Dugway Proving Ground	Tooele, UT
6	*Hill Field	
7	Logan	Utah ASF Depot
8	*Ogden	
9	Orem	
10	Salina	Utah ASF Depot
11	Tooele	
12	Tremonton	Utah ASF Depot
13	Utah ASF Depot	
14	Warren	

Bushnell General Hospital

The POW camp was located near Bushnell General Hospital. This facility was located in Brigham City, UT. It opened for POWs in February 1944 and closed in May 1946. Some of the German and Italian POWS worked in the hospital, under close supervision.

Naval Supply Depot Clearfield

The depot was opened in April 1943 as the major Naval supply depot. In November 1943, Italian POWs were put to work in the depot to alleviate the wartime shortage of manpower. Later, 500 German POWs were at this camp. The last POWs were returned to their homeland, and the camp closed in May 1946.

Desert Chemical Warfare Depot

This POW Camp opened in November 1944 and closed in October 1945. The camp held 108 POWs.

Ft. Douglas

Fort Douglas received 250 Italian prisoners of war in October of 1944. The Italians generally gave little trouble to the post. Some old CCC buildings were used to house the prisoners for their stay at Fort Douglas. With the end of the war in September of

1945, the Italian prisoners were sent back to their homeland and replaced by some German prisoners. The German prisoners were used to harvest sugar beets at the urgent request of farmers in the area. They were not repatriated until July of 1946.

Hill Field

Hill Air Force Base (HAFB), originally known as Hill Field, was the site for the Ogden Air Depot, whose name underwent several changes until it became the Ogden Air Material Area. It was located five miles to the south of Ogden, Utah. The POW camp opened in March 1944 and closed in May 1946. There were 755 German and Italian POWs housed there.

Ogden

During World War II, Ogden was the largest quartermaster depot in the United States. It employed 4,000 civilians and 5,000 German and Italian prisoners of war.

UT-8-1-1

UT-8-1-5

UT-8-1-10

No.	Denomination	Paper	Ink	Serial	Block	Circ.	Unc.
UT-8-1-1	1 cent	tan	black	red		150	250
UT-8-1-5	5 cents	tan	black	red		150	250
UT-8-1-10	10 cents	rose	black	red		150	250

Toole

The Toole Army Depot was built during the war because the Ogden Arsenal, with increasing responsibilities, could not physically expand. In October 1944 the POW camp opened. The prisoner of war camp was located on the north side of the maintenance and storage area of the depot. It consisted of 29 barracks, which were sold as war surplus following World War II. The POW camp closed in May of 1946. Since 1949, the prisoner of war camp site has been used by the salvage department.

Utah ASF

Ogden ASF Depot was later known as Defense Depot Ogden (DDOU). The POW camp opened in March 1944 and closed July 1946. German POWs interned at the Utah ASF Depot Prisoner of War Camp, were assigned to work at the Utah Poultry Producers Cooperative Association Plant at Tremonton, Utah. Each man was given one specific job to do.

VIRGINIA

	Camp	Base Camp
1	Allen	
2	*Camp Ashby	
3	*Camp Ashby Branch Camp	
4	*Camp Patrick Henry	
5	*Camp Peary	
6	*Camp Pickett	
7	Catawba	Pickett, VA
8	Cheatam Annex	Peary, VA
9	Cumberland	Pickett, VA
10	*Ettenger	Somerset, MD
11	Fairfax	Front Royal, VA
12	*Fort. Eustis	
13	Fort. Eustis Special Projects	
14	*Front Royal	
15	Green Bay	Pickett, VA
16	Hampton Roads POE	
17	Lee	
18	Leesburg	Front Royal
19	Little Creek	Ashby, VA
20	Lyndhurst	Pickett, VA
21	Naval Armed GS	Ashby
22	Radford	Pickett, VA
23	Richmond AAB	
24	Richmond ASF	
25	Salem	Pickett, VA

26	Sandy Level	Pickett, VA
27	Shelton	Pickett, VA
28	Suffolk	Ashby, VA
29	Timberville	Pickett, VA
30	White Hall	Pickett, VA
31	Winchester	Front Royal, VA

Camp Allen Naval Operating Base

This camp was located in Norfolk, VA. The first POWs arrived in May 1945 and were sent home in May of 1946.

Camp Ashby

Camp Ashby was located on a 200 acre farm in Princess Anne County-South Hampton Roads. The camp opened in March of 1944 and closed in May of 1946. In its time, more than 6,000 German POWs were encamped there.

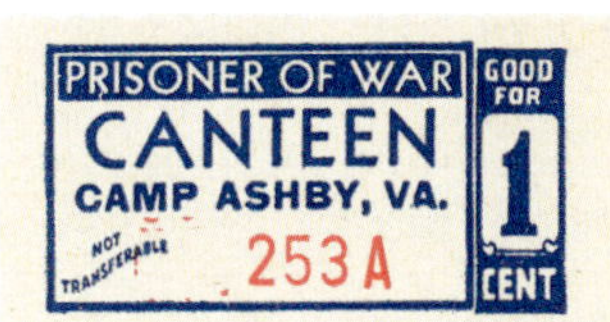

VA-2-1-1

VA-2-1-5

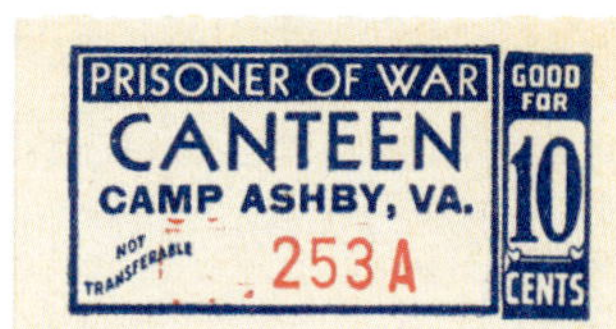

VA-2-1-10

No.	Denomination	Paper	Ink	Serial	Block	Circ.	Unc.
VA-2-1-1	1 cent	white	blue	red	A	25	40
VA-2-1-5	5 cents	white	blue	red	A	35	55
VA-2-1-10	10 cents	white	blue	red	A	55	85

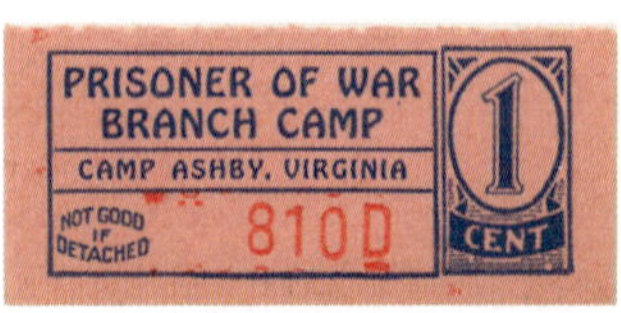

VA-3-1-1

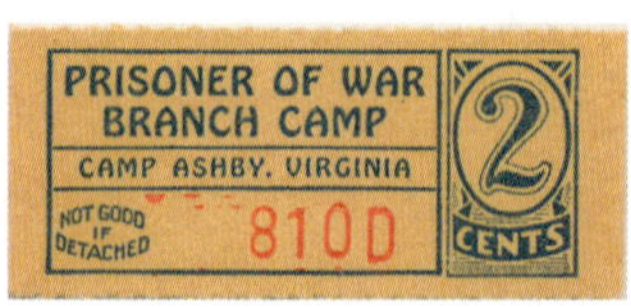

VA-3-1-2

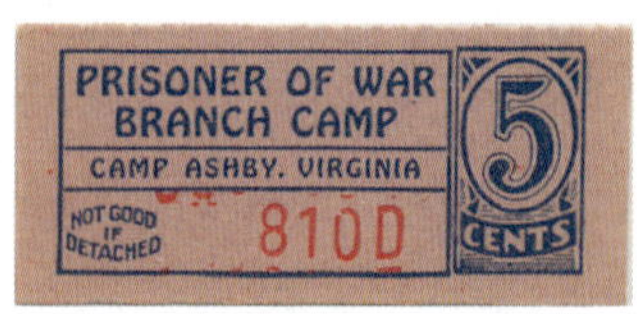

VA-3-1-5

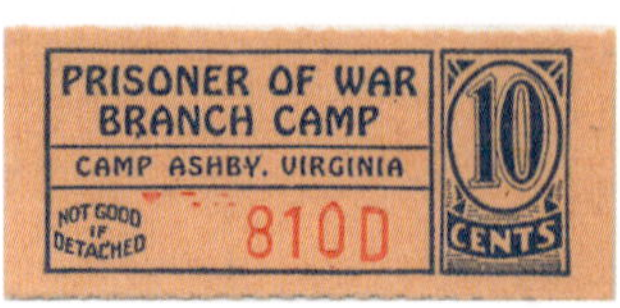

VA-3-1-10

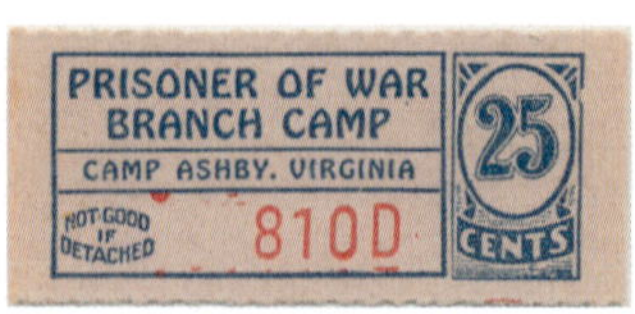

VA-3-1-25

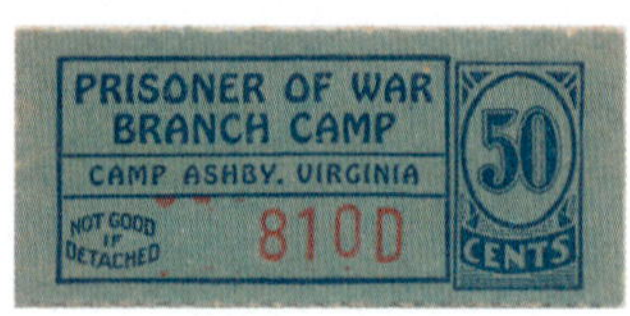

VA-3-1-50a

VA-3-1-50b

No.	Denomination	Paper	Ink	Serial	Block	Circ.	Unc.
VA-3-1-1	1 cent	pink	blue	red	D	50	100
VA-3-1-2	2 cents	tan	blue	red	D	75	150
VA-3-1-5	5 cents	mauve	blue	red	D	50	100
VA-3-1-10	10 cents	peach	blue	red	D	50	100
VA-3-1-25	25 cents	gray	blue	red	D	50	100
VA-3-1-50a	50 cents	green	blue	red	D	75	150
VA-3-2-50b	50 cents	gray	red	red	H	150	250

Camp Ettinger

Camp Ettinger was located in Northampton county. The camp opened in October, 1944 and closed in May of 1946. It had a population of 6,580 POWs. This was not a base camp.

Camp Front Royal Remount Depot

The first POWs were captured Russian soldiers working on the "West Wall." Later internees were captured German POWs from Africa. The camp opened in July 1944 and closed in May 1946.

VA-14-1-1

No.	Denomination	Paper	Ink	Serial	Block	Circ.	Unc.
VA-14-1-1	1 cent	pink	black	red		150	250

Ft. Eustis

At the end of the war, there were still 4,100 Germans and 1,300 Italians on the Peninsula. Almost totally forgotten is the fact that a secret experiment took place at Fort Eustis in 1945/1946 to re-educate Nazi prisoners. The purpose of this was to create a core of cooperative and pro-American Germans to be repatriated and then help rebuild in the American zone of occupied Germany. In all, 20,000 POWs from all over the United States were processed through a six-day course at Fort Eustis before returning home!

VA-12-1-1

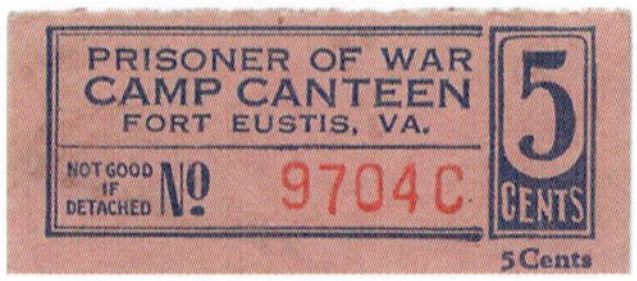

VA-12-1-5

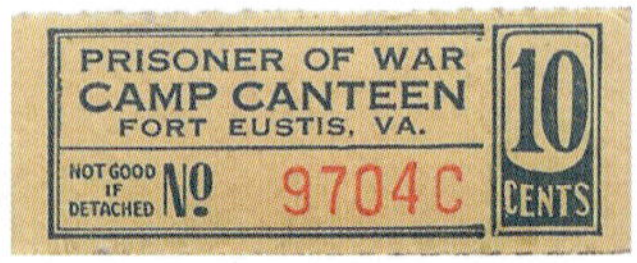

VA-12-1-10

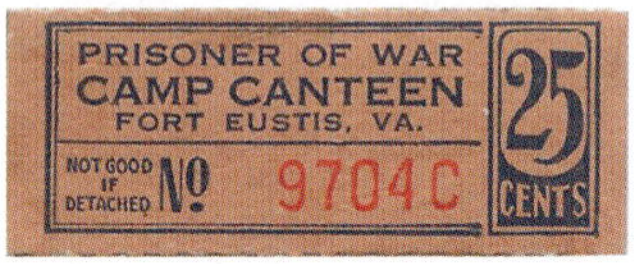

VA-12-1-25

No.	Denomination	Paper	Ink	Serial	Block	Circ.	Unc.
VA-12-1-1	1 cent	cream	blue	red	C	100	150
VA-12-1-5	5 cents	rose	blue	red	C	125	175
VA-12-1-10	10 cents	yellow	blue	red	C	150	200
VA-12-1-25	25 cents	brown	blue	red	C	175	225

Hampton Roads POE

This was the main port of entry for the majority of the German and Italian POWS during WWII.

Camp Hill, Newport News, VA

Camp Hill, also used for the Italian POWs, was bounded to the south by the temporary wooden railroad overpass at 58th Street, the James River Bridge/Military Highway railroad overpass to the north, Jefferson Avenue to the east, and the railroad yards to the west. After Italy switched sides in the war, more than a thousand Italian POWs housed in Camp Hill (not to be confused with the much larger Camp A.P. Hill, near Fredericksburg), were re-designated as 'Italian Service Unit Personnel' and employed in the port area. Hundreds more worked on farms in Warwick County and in 46 businesses (not identified) on the peninsula.

The most visible, to the public, of the prisoner enclosures was a barbed wire enclosed area on the north side. Although isolated from the rest of the military complex, anyone riding over the railroad overpass could literally look down into that encampment and "see the enemy."

Camp Patrick Henry

This camp opened in February 1944 and closed in March 1946. During that time, 556 POWS were held in the camp.

VA-4-1-1a

VA-4-1-1b

VA-4-1-1c

VA-4-1-1d

VA-4-1-5a

VA-4-1-5b

VA-4-1-5c

VA-4-1-10a

VA-4-1-10b

VA-4-1-10c

VA-4-1-25a

VA-4-1-25b

No.	Denomination	Paper	Ink	Serial	Block	Circ.	Unc.
VA-4-1-1a	1 cent	gray	blue	red	A	25	40
VA-4-1-1b	1 cent	green	blue	red	A	25	40
VA-4-1-1c	1 cent	cream	blue	red	C	25	40
VA-4-1-1d	1 cent	peach	red	red	G	25	40
VA-4-1-5a	5 cents	white	blue	red	A,C	35	55
VA-4-1-5b	5 cents	orange	blue	red	C	35	55
VA-4-1-5c	5 cents	rose	red	red	G	35	55
VA-4-1-10a	10 cents	gray	blue	red	C	55	85
VA-4-1-10b	10 cents	cream	blue	red	A,C	55	85
VA-4-1-10c	10 cents	pink	red	red	G	55	85
VA-4-1-25a	25 cents	peach	blue	red	C	90	125
VA-4-1-25b	25 cents	pink	blue	red	C	90	125

Camp Peary

The POWs kept at Camp Peary were not just an ordinary sort, but, rather, many came from captured German submarine and ship crews that the Germans had thought were lost at sea, the crews presumed dead. It was important for Nazi authorities to be unaware of their capture, since that also meant secret code books, presumed lost at sea, may also have been compromised. Thus, extra secrecy was necessary.

Many of the former POWs stayed in Virginia and the United States after the war and became naturalized as U.S. citizens.

VA-5-1-5

No.	Denomination	Paper	Ink	Serial	Block	Circ.	Unc.
VA-5-1-5	5 cents	gray	blue	red	F	150	250

Camp Pickett

This camp was located just east of the town of Blackstone. The camp opened in February 1944 and closed in May of 1946. The main camp held 2,676 POWS, with an additional 3,300 in various sub-camps.

VA-6-1-1

VA-6-1-5

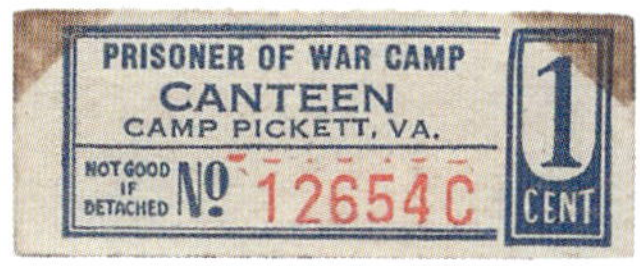

VA-6-2-1

No.	Denomination	Paper	Ink	Serial	Block	Circ.	Unc.
	Not Good if Detatched below SN						
VA-6-1-1	1 cent	pink	black	red		150	250
VA-6-1-5	5 cents	tan	black	red		150	250
	Not Good if Detached on same line						
VA-6-2-1	1 cent	white	blue	red	C	100	200

Richmond AAB

Richmond Army Air Base, Richmond, Chesterfield County, VA (base camp) was opened in May 1945, and held 150 POWs. The POW Camp closed in September 1945.

Richmond ASF

Richmond Armed Service Forces Depot held 1,311 POWs in the Chesterfield County camp from January 1945 until May 1946.

WASHINGTON

	Camp	Base Camp
1	Barnes General Hospital	Ft. Lewis, WA
2	Baxter General Hospital	Ft. Lewis, WA
3	*Ft. Lawton	
4	*Ft. Lewis	
5	Malott	Ft. Lewis, WA
6	McCaw General Hospital	Ft. Lewis, WA
7	Peshastin	Ft. Lewis, WA
8	Toppenish	Ft. Lewis, WA
9	Walla Walla	Ft. Lewis, WA
10	Vancouver Barracks	Ft. Lewis, WA

Ft. Lawton

The POW camp opened March 1945 and closed in June 1946. A total of 897 POWs were housed there. Up to 5,000 Italian POWs passed through the camp on route to Hawaii for internment.

WA-3-1-5

WA-3-1-10

WA-3-1-25

No.	Denomination	Paper	Ink	Serial	Block	Circ.	Unc.
WA-3-1-5	5 cents	red	black	red		200	300
WA-3-1-10	10 cents	tan	black	red		200	300
WA-3-1-25	25 cents	yellow	black	red		200	300

Ft. Lewis

In early 1942 the first POWs arrived; four Japanese, two Italians, and a German. They were quickly transferred to other camps. At its peak, 4,245 German POWs were held at the camp. The camp closed in June 1946.

WA-4-1-1

WA-4-1-2

WA-4-1-5

WA-4-1-10

WA-4-2-1

WA-4-2-2

WA-4-2-5

WA-4-2-10

WA-4-2-25

WA-4-3-1

WA-4-3-2

WA-4-3-5

WA-4-3-10

WA-4-3-25

No.	Denomination	Paper	Ink	Serial	Block	Circ.	Unc.
	Diamond Border						
WA-4-1-1	1 cent	blue	black	red		75	125
WA-4-1-2	2 cents	green	black	red		100	150
WA-4-1-5	5 cents	orange	black	red		100	150
WA-4-1-10	10 cents	brown	black	red		125	175
	Zigzag Border						
WA-4-2-1	1 cent	blue	black	red		75	125
WA-4-2-2	2 cents	green	black	red		100	150

WA-4-2-5	5 cents	orange	black	red		100	150
WA-4-2-10	10 cents	brown	black	red		125	175
WA-4-2-25	25 cents	yellow	black	red		150	200
	Canteen on Top Line Red PW						
WA-4-3-1	1 cent	blue	black	red	A	75	125
WA-4-3-2	2 cents	green	black	red	A	100	150
WA-4-3-5	5 cents	orange	black	red	A	100	150
WA-4-3-10	10 cents	brown	black	red	A	125	175
WA-4-3-25	25 cents	yellow	black	red	A	150	200

POW Canteen, Ft. Lewis, Washington

WEST VIRGINIA

	Camp
1	Ashford General Hospital
2	Baker General Hospital
3	Rainelle
4	*White Springs General Hospital

Ashford General Hospital

The 350 Italian POWs arrived in the summer of 1942 to build Camp Ashford, a standard U.S. Army double-barbed wire compound, near White Sulphur Springs. The Italian POWs were moved, and 679 German POWs arrived to maintain the grounds and work in the hospital. The POW camp closed in June 1946.

White Sulphur Springs

WV-4-1-10

No.	Denomination	Paper	Ink	Serial	Block	Circ.	Unc.
WV-4-1-10	10 cents	rose	blue	red	B	200	300

WISCONSIN

	Camp	Base Camp
1	Antigo	Sheridan, IL
2	Appleton	Sheridan, IL
3	Barron	Sheridan, IL
4	Bayfield	Sheridan, IL
5	Cambria	Sheridan, IL
6	*Camp McCoy	
7	Chilton	Sheridan, IL
8	Cobb	Sheridan, IL

9	Columbus	Sheridan, Il
10	Fairgrounds	Sheridan, Il
11	Fish Creek	Sheridan, IL
12	Fox Lake	Sheridan, Il
13	Freedonia	Sheridan, Il
14	Friedlund's Orchards	Sheridan, IL
15	Galesville	Sheridan, IL
16	Genessee Depot	Sheridan, IL
17	Goldman's Orchards	Sheridan, IL
18	Hartford	Billy Mitchell Field
19	Hortonville	Sheridan, IL
20	Janesville	Sheridan, IL
21	Lodi	Sheridan, IL
22	Markesan	Sheridan, Il
23	Marshfield	Sheridan, Il
24	Miller's Orchards	Sheridan, IL
25	Milltown	Sheridan, IL
26	*Billy Mitchell Field	
27	Oakfield	Sheridan, IL
28	Plymouth	Sheridan, IL
29	Reedsburg	Sheridan, IL
30	Reynolds' Orchards	Sheridan, IL
31	Rhinelander	Sheridan, IL
32	Ripon	Sheridan, IL
33	Rockford	Billy Mitchell Field
34	Sheboygan	Sheridan, IL
35	Sturgeon Bay	Sheridan, IL
36	Sturtevant	Sheridan, IL
37	Waterloo	Sheridan, IL
38	Waupun	Sheridan, IL
39	Wisconsin Rapids	Sheridan, IL

Billy Mitchell Field

On January 4, 1945, Mitchell Field was leased to the War Department as a POW camp. There were 776 German POWs held here. The camp closed in May 1946.

WI-5-1-1

WI-5-1-5

WI-5-1-10

No.	Denomination	Paper	Ink	Serial	Block	Circ.	Unc.
WI-5-1-1	1 cent	cream	blue	red	A	150	250
WI-5-1-5	5 cents	orange	blue	red	A	150	250
WI-5-1-10	10 cents	tan	blue	red	A	150	250

Camp McCoy

Located in Monroe county, the camp opened in June 1943 and closed in May 1946. The War Department announced that every Japanese prisoner of war brought to the United States, a total of 62, was interned at the Camp McCoy Center. The camp was the largest holding facility for Japanese POWs in the Continental United States. Camp McCoy is unique in American history as having housed relocated Japanese-Americans from the West Coast, 290 German POWs, and the 62 Japanese prisoners of war captured during World War II.

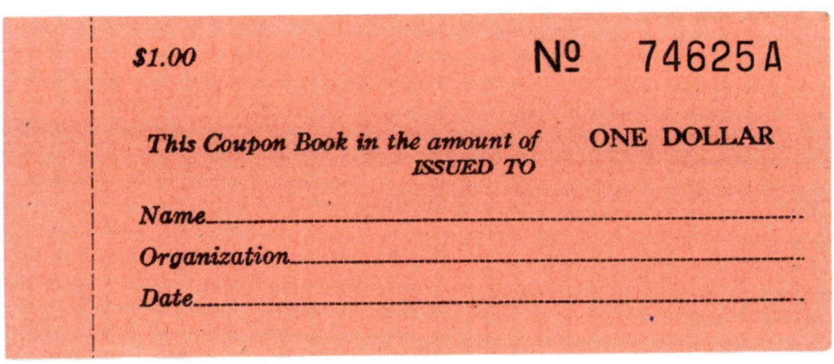

WI-6-1-1a

WI-6-1-1b

WI-6-1-5

WI-6-1-10

WI-6-2-1

WI-6-2-5 back

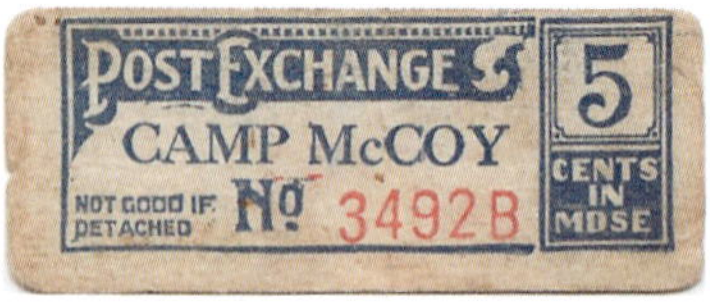

WI-6-2-5

No.	Denomination	Paper	Ink	Serial	Block	Circ.	Unc.
WI-6-1-1a	1 cent	cream	blue	red	D	150	250
WI-6-1-1b	1 cent	green	blue	red	A	150	250
WI-6-1-5	5 cents	yellow	blue	red	A	150	250
WI-6-1-10	10 cents	orange	blue	red	A	150	250
	Exchange						
WI-6-2-1	1 cent	cream	blue	red	A,B	150	250
WI-6-2-5	5 cents	cream	blue	red	B	150	250

WYOMING

	Camp	Base Camp
1	Basin	Scottsbluff, NE
2	Centennial	Greeley, CO
3	Clearmont	Douglas, WY
4	*Douglas	
5	Douglas II	Scottsbluff, NE
6	DuBois	Scottsbluff, NE
7	Easterbrook	Douglas. WY
8	*Fort F.E. Warren	
9	Lingle	Scottsbluff, NE
10	Lovell	Douglas, WY
11	Pine Bluff	Scottsbluff, NE
12	Powell	Douglas, WY
13	Riverton	Douglas, WY
14	Ryan Park	Greeley, CO
15	Saratoga	Douglas, WY
16	Torrington	Scottsbluff, NE
17	Veteran	Scottsbluff, NE
18	Wheatland	Douglas, WY
19	Wind River	Scottsbluf, NE
20	Worland	Douglas, WY

Douglas

In August 1942, the first 412 Italian POWs arrived. The 1,900 Italian POWS were sent to ISU camps in 1944 when Italy surrendered, and the camp was deactivated in July 1944.

The camp was reactivated and, in October, 2,000 German POWs arrived. The camp closed in December 1945. The camp is the site for a number of murals by three Italian POWs. The murals, some 8 feet by 15 feet, are in the former Officers Club.

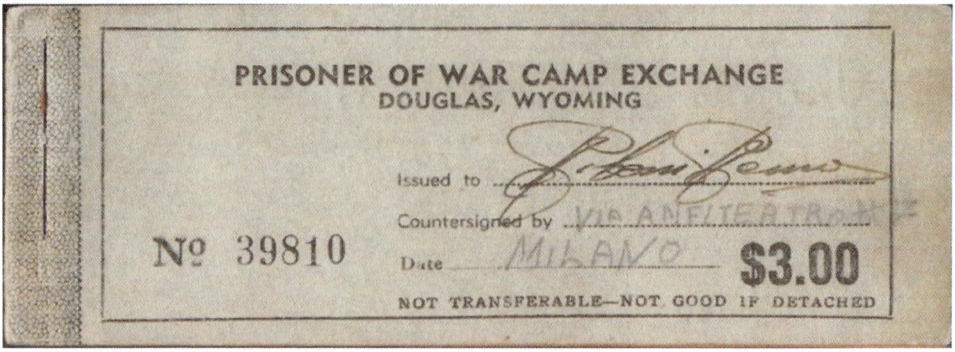

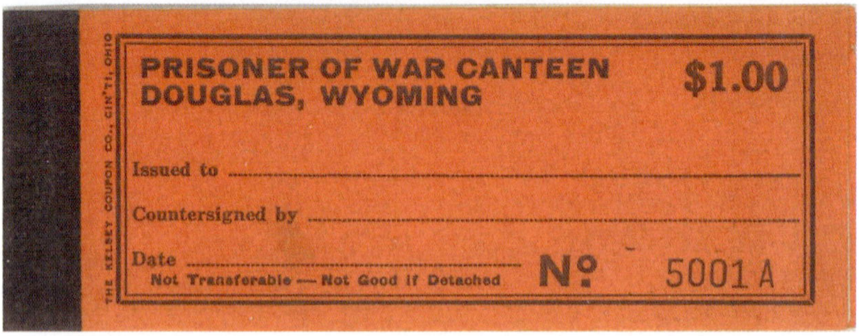

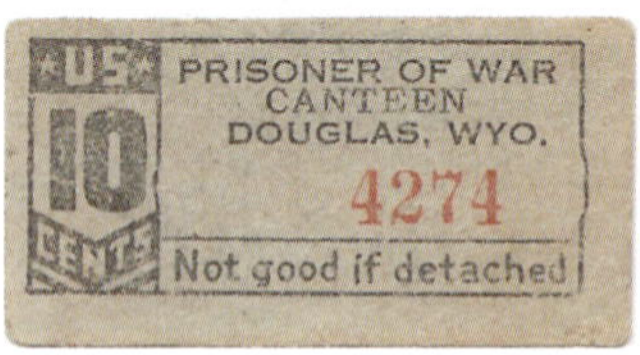

WY-4-1-10

WY-4-2-1a

WY-4-2-1b

WY-4-2-5

WY-4-2-10

No.	Denomination	Paper	Ink	Serial	Block	Circ.	Unc.
	Denomination on Left						
WY-4-1-10	10 cents	gray	black	red		150	250
	Denomination on Right						
WY-4-2-1a	1 cent	orange	blue	red	A	25	40
WY-4-2-1b	1 cent	orange	red	red	A,B	25	40
WY-4-2-5	5 cents	yellow	blue	red	A	35	55
WY-4-2-10	10 cents	gray	blue	red	A	55	85

Ft. F.E. Warren

Fort Francis E. Warren was located outside of Casper. On November 12, 1943, the first group of 350 German POWs, from Rommel's Afrika Korps, arrived. The population climbed to 546 before the camp closed in May 1946.

No.	Denomination	Paper	Ink	Serial	Block	Circ.	Unc.
WY-8-1-1	1 cent	pink	blue	red		150	250
WY-8-1-5	5 cents	tan	blue	red		150	250
WY-8-1-10	10 cents	pink	blue	red		150	250
WY-8-1-25	25 cents	peach	blue	red		150	250

The last batch of POWs leaving the United States

Italian POWs doing road work saluting a U.S. officer

INTERNMENT CAMPS

During World War II, internment camps were set up for resident aliens from the Axis Powers. South American countries deported many of their resident Axis aliens to the USA for internment. The internment camps did not have any residents that were named under Executive Order 9066, operated by the War Relocation Authority. The internment camps were set up under the authority of the Immigration and Naturalization Service.

The camps held German, Italian, and Japanese citizens from the USA. In addition, many German, Italian, and Japanese from South America were deported to the USA for internment, primarily to camps in Texas.

The camps issued paper chits in booklet form, just like the Italian Service Units and the POW camps. Of the three types of camps in the USA for World War II, the Internment camps are the scarcest.

While the internment camps issued paper chits, as well as several token issues, our intent is to cover only the paper chits similar to the POW chits. There were nonmetallic I. & N. Service general tokens issued for camps at Crystal City and Seagoville, TX.

State	Internment Camp	Camp Number
Alabama	Opelika	1
Arizona	Florence	2
California	Angel Island	3
	Griffith Park	4
Colorado	Carson	5
	Trinidad	6
Florida	Blanding	7
Georgia	Ft. Oglethorpe	8
Indiana	Atterbury	9
Louisiana	Livingston	10
Maryland	Ft. Meade	11
Massachusetts	Ft. Devens	12
Mississippi	Como	13
	McCain	14
	Shelby	15
Missouri	Clark	16
	Ft. Leonard Wood	17
	Weingarten	18

Nebraska	Ft. Robinson	19
New Mexico	Lordsburg	20
	Roswell	21
New York	Upton	22
North Carolina	Ft. Bragg	23
Oklahoma	Alva	24
	McAlester	25
	Ft. Sill	26
	Stringtown	27
Tennessee	Crossville	28
	Forrest	29
Texas	Ft. Bliss	30
	Ft. Sam Houston	31
	Huntsville	32
	McLean	33
	Mexia	34
Washington	Ft. Lewis	35
Wisconsin	McCoy	36

Alva, Oklahoma Internment Camp

OKI-24-1-1

OKI-24-1-5

OKI-24-1-10

No.	Denomination	Paper	Ink	Serial	Block	Circ.	Unc.
OKI-24-1-1	1 cent	blue	blue	red	B	25	40
OKI-24-1-5	5 cents	tan	blue	red	B	35	55
OKI-24-1-10	10 cents	rose	blue	red	B	55	85

Atterbury, Indiana Internment Camp

INI-9-1-1

INI-9-1-5

INI-9-1-10

No.	Denomination	Paper	Ink	Serial	Block	Circ.	Unc.
INI-9-1-1	1 cent	blue	blue	red	C	25	40
INI-9-1-5	5 cent	yellow	blue	red	C	35	55
INI-9-1-10	10 cent	orange	blue	red	C	55	85

Camp Carson, Colorado Internment Camp

COI-5-1-5a

COI-5-1-5b

COI-5-1-10

No.	Denomination	Paper	Ink	Serial	Block	Circ.	Unc.
COI-5-1-5a	5 cents	pink	black	red		100	225
COI-5-1-5b	5 cents	brown	black	red		100	225
COI-5-1-10	10 cents	blue	black	red		150	300

Camp Clark, Missouri Internment Camp

MOI-16-1-1

MOI-16-1-5

MOI-16-1-10

MOI-16-1-25

No.	Denomination	Paper	Ink	Serial	Block	Circ.	Unc.
MOI-16-1-1	1 cent	pink	blue	red	A,C	25	40
MOI-16-1-5	5 cents	peach	blue	red	A,C	35	55
MOI-16-1-10	10 cents	yellow	blue	red	A,C	55	85
MOI-16-1-25	25 cents	brown	blue	red	A,C	90	150

Como, Tennessee Internment Camp

MSI-13-1-1

MSI-13-1-5

MSI-13-1-25

No.	Denomination	Paper	Ink	Serial	Block	Circ.	Unc.
MSI-13-1-1	1 cent	pink	blue	red	A,C,E	25	40
MSI-13-1-5	5 cents	peach	blue	red	A,C,E	35	55

MSI-13-1-10	10 cents	yellow	blue	red	A,C,E	55	85
MSI-13-1-25	25 cents	brown	blue	red	C,E	90	150

Crossville, Tennessee Internment Camp

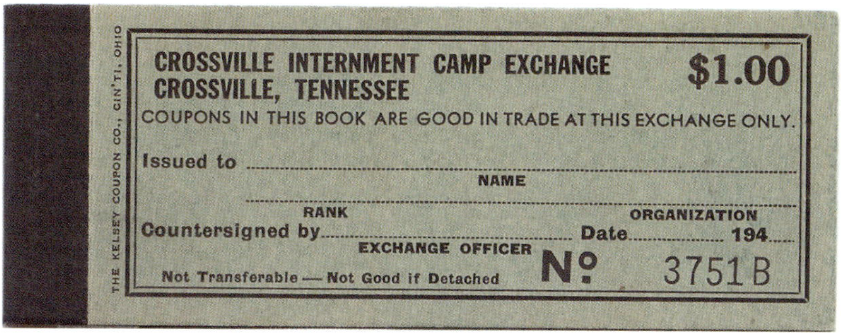

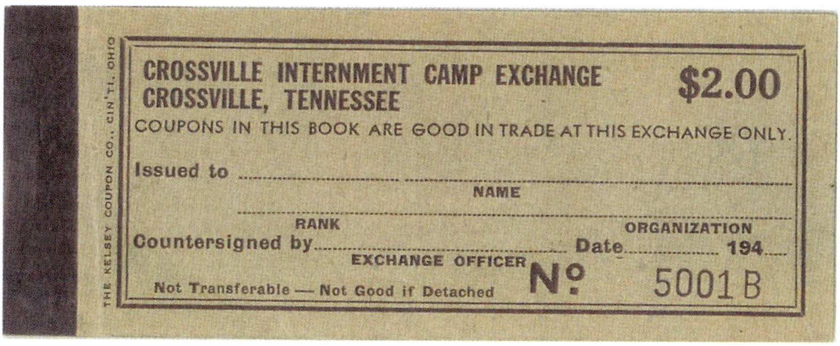

TNI-28-1-1

TNI-28-1-5

TNI-28-1-10

TNI-28-2-1

TNI-28-2-5

TNI-28-2-10

No.	Denomination	Paper	Ink	Serial	Block	Circ.	Unc.
TNI-28-1-1	1 cent	blue	blue	red	B	25	40
TNI-28-1-5	5 cents	peach	blue	red	B	35	55
TNI-28-1-10	10 cents	gray	blue	red	B	55	85

	Exchange/						
TNI-28-2-1	1 cent	blue	blue	red	B	25	40
TNI-28-2-5	5 cents	peach	blue	red	B	35	55
TNI-28-2-10	10 cents	gray	blue	red	B	55	85

Fort Robinson, Nebraska Internment Camp

No.	Denomination	Paper	Ink	Serial	Block	Circ.	Unc.
NEI-19-1-1	1 cent	grey	blue	red	A,B,C	40	75
NEI-19-1-5	5 cents	buff	blue	red	A,B,C	80	125
NEI-19-1-10	10 cents	peach	blue	red	A,B,C	135	175
NEI-19-1-25	25 cents	green	blue	red	C,E		RNK

Livingston, Louisiana Internment Camp

LAI-10-1-1

No.	Denomination	Paper	Ink	Serial	Block	Circ.	Unc.
LAI-10-1-1	1 cent	peach	black	black	A	175	350

McLean, Texas Internment Camp

TXI-33-1-1

No.	Denomination	Paper	Ink	Serial	Block	Circ.	Unc.
TXI-33-1-1	1 cent	gray	black	red		150	250

Mexia, Texas Internment Camp

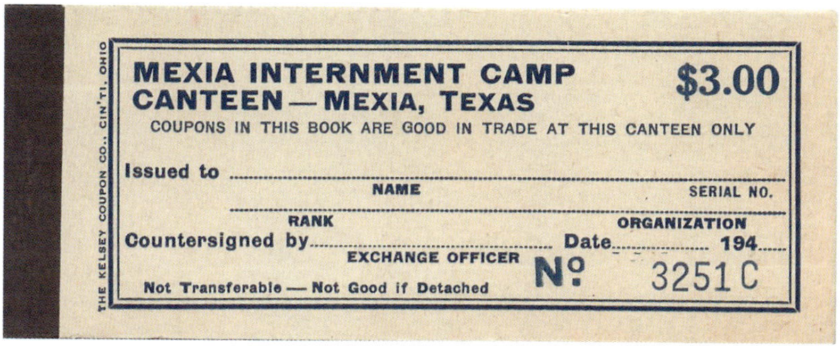

TXI-33-1-1

TXI-33-1-5

TXI-33-1-10

TXI-33-1-25

No.	Denomination	Paper	Ink	Serial	Block	Circ.	Unc.
TXI-33-1-1	1 cent	orange	black	red	A,B,C	25	40
TXI-33-1-5	5 cents	yellow	black	red	A,B,C	35	55
TXI-33-1-10	10 cents	gray	black	red	A,B,C	55	85
TXI-33-1-25	25 cents	rose	black	red	C	90	150

Opelika, Alabama Internment Camp

ALI-1-1-1a

ALI-1-1-1b

ALI-1-1-5

No.	Denomination	Paper	Ink	Serial	Block	Circ.	Unc.
ALI-1-1-1a	1 cent	brown	blue	red	B	100	200
ALI-1-1-1b	1 cent	blue	blue	red	B	100	200
ALI-1-1-5	5 cents	white	blue	red	B	150	300
ALI-1-1-10	10 cents	red	blue	red	B	175	350

Roswell, New Mexico Internment Camp

NMI-21-1-1

NMI-21-1-5

NMI-21-1-10

No.	Denomination	Paper	Ink	Serial	Block	Circ.	Unc.
NMI-21-1-1	1 cent	orange	blue	red	C	25	40
NMI-21-1-5	5 cents	pink	blue	red	C	35	55
NMI-21-1-10	10 cents	tan	blue	red	C	55	85
NMI-21-1-25	25 cents	brown	blue	red	C	90	150

Trinidad, Colorado Internment Camp

COI-6-1-1

No.	Denomination	Paper	Ink	Serial	Block	Circ.	Unc.
COI-6-1-1	1 cent	yellow	black	red		175	350

Weingarten, Missouri Internment Camp

MOI-18-1-1

MOI-18-1-5

MOI-18-1-10

No.	Denomination	Paper	Ink	Serial	Block	Circ.	Unc.
MOI-18-1-1	1 cent	peach	black/red	red		150	300
MOI-18-1-5	5 cents	tan	black/red	red		175	350
MOI-18-1-10	10 cents	brown	black/red	red		200	400

ITALIAN SERVICE UNITS

Of the Italian Prisoners of War that were captured during World War II, about 50 thousand were transported to the USA. They were kept in the many POW camps in America. On September 8, 1943, with the signing of the armistice, Italy became an ally of the Allied Powers.

The USA still had a critical shortage of manpower across many war related labor needs. A partial list includes food production, domestic transportation of war materials for the European and Pacific theaters, and all manner of shipping port jobs. It was determined that the former Italian POWs, now part of the co-belligerent forces, could take on a greater role, filling in the labor shortages that POWs would not be assigned due to the sabotage concerns.

In February 1944, the Italian Service Units were established. The Italian POWs that agreed to work for the war effort were quickly transported to coastal and industrial sites. While hard numbers are not available, the best guess at participation in the program was 90%. They were housed in separate locations under different conditions. They were afforded increased responsibilities and given the opportunity to interact with Americans and see the country.

They were paid for their labor in paper chits, in booklets similar to what they had as POWs. There are four camps for which we have examples of these chits. The three New Jersey locations have "Italian Service Branch" printed on their chits. The Port Terminal NYPE Exchange have "CUPONI STACCATI NON SONO VALIDI" written on the chits. The translation is the same message printed on essentially all POW chits, "not good if detached;" however, "Italian Service Branch" is not written on the chits.

An asterisk next to the name in the chart below indicates the ISU issued chits.

	Italian Service Units
	California
1	Benicia Ars.
2	Camp Stoneman
3	Fort McArthur
4	Fort McDowell
5	Fort Ord
6	Haan
7	Irwin
8	J. T. Knight
9	Lathrop Eng'r Depot

10	Mira Lomia QMSD
11	Presidio of San Fran
12	Roberts
13	Ross LAPoE
14	San Bernardino Eng'r Depot
15	San Luis Obispo
16	Yermo H&R
	Georgia
17	Fort Benning
	Illinois
18	Savanna Ord. Depot Prov Gr
19	Vaughn General Hospital
	Kentucky
20	Fort Knox
	Louisiana
21	Jackson Bks. NO POE
	Maryland
22	Fort George G. Meade
	Massachusetts
23	Fort Andrews Boston
	Michigan
24	Fort Wayne Ord. Depot
	Nebraska
25	Sioux Ord Depot

	New Jersey
26	*Belle Mead
27	Fort Dix
28	Fort Mounmouth
29	*Kilmer
30	*Raritan Ars. (Metuchen)
	New York
31	Elmira H & R Horseheads
32	*Fort Hamilton NYPE
33	Fort Jay
34	Shanks
35	Staten Island Terminal Stapleton
	North Carolina
36	Fort Bragg
37	Fort Butner
	Ohio
38	Erie Proving Grounds La Carmem
	Pennsylvania
39	Letterkenny Ord. Depot Chambg.
	Utah
40	Desert Center Chem Warfare Dpt
41	Dugway Proving Ground
42	Tooele Ord. Depot
43	Utah A.S.F. Depot (Ogden)
	Virginia
44	Fort Patrick Henry

45	Hampton Roads POE Newport N
46	Norfolk Army Base
	Washington
47	Mt. Rainer Ord. Depot
48	Seattle A.S.F. Depot
49	Vancouver Barracks

Belle Mead ISU

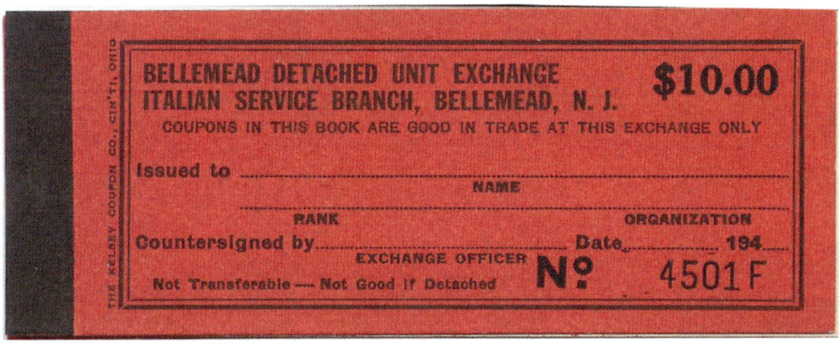

NJU-26-1-1

NJU-26-1-5

NJU-26-1-10

NJU-26-1-25

NJU-26-1-50

No.	Denomination	Paper	Ink	Serial	Block	Circ.	Unc.
NJU-26-1-1	1 cent	white	blue	red	F	25	40
NJU-26-1-5	5 cents	rose	blue	red	F	35	55
NJU-26-1-10	10 cents	buff	blue	red	F	55	85
NJU-26-1-25	25 cents	gray	blue	red	F	90	150
NJU-26-1-50	50 cents	green	blue	red	F	125	200

Kilmer, New Jersey ISU

NJU-29-1-1

NJU-29-1-5

NJU-29-1-10

NJU-29-1-25

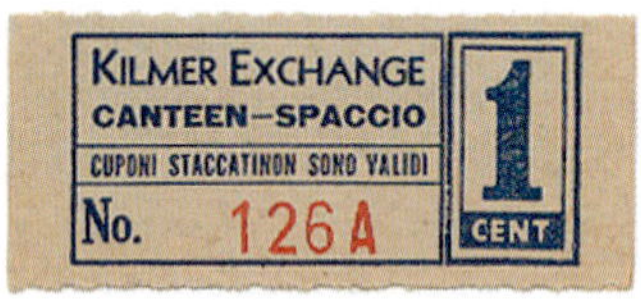

NJU-29-2-1

NJU-29-2-10

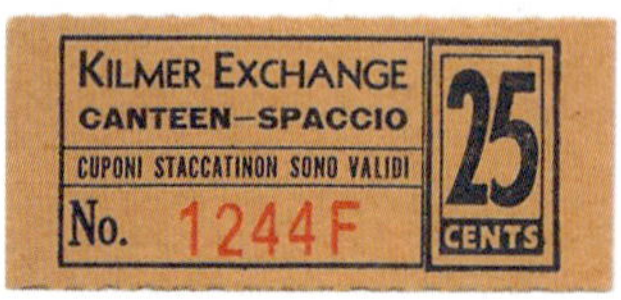

NJU-29-2-25

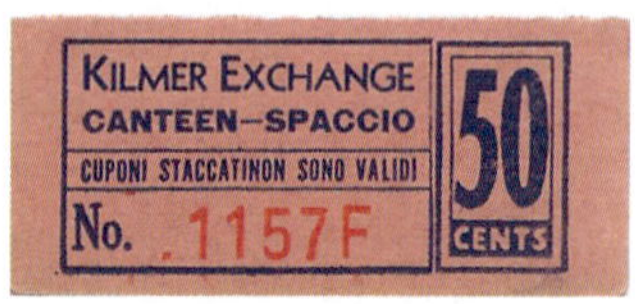

NJU-29-2-50

No.	Denomination	Paper	Ink	Serial	Block	Circ.	Unc
	Exchange						
NJU-29-1-1	1 cent	white	green	red	1 B	25	40
NJU-29-1-5	5 cents	pale orange	green	red	1 B	35	55
NJU-29-1-10	10 cents	orange	green	red	1 B	55	85
NJU-29-1-25	25 cents	gray	green	red	1 B	90	150
	Canteen Spaccio						
NJU-29-2-1	1 cent	gray	blue	red	A,F	25	40
NJU-29-2-5	5 cents	pale orange	blue	red	A,F	35	55
NJU-29-2-10	10 cents	pale green	blue	red	A,F	55	85
NJU-29-2-25	25 cents	pale yellow	blue	red	A,F	90	150
NJU-29-2-50	50 cents	pink	blue	red	A,F	125	200

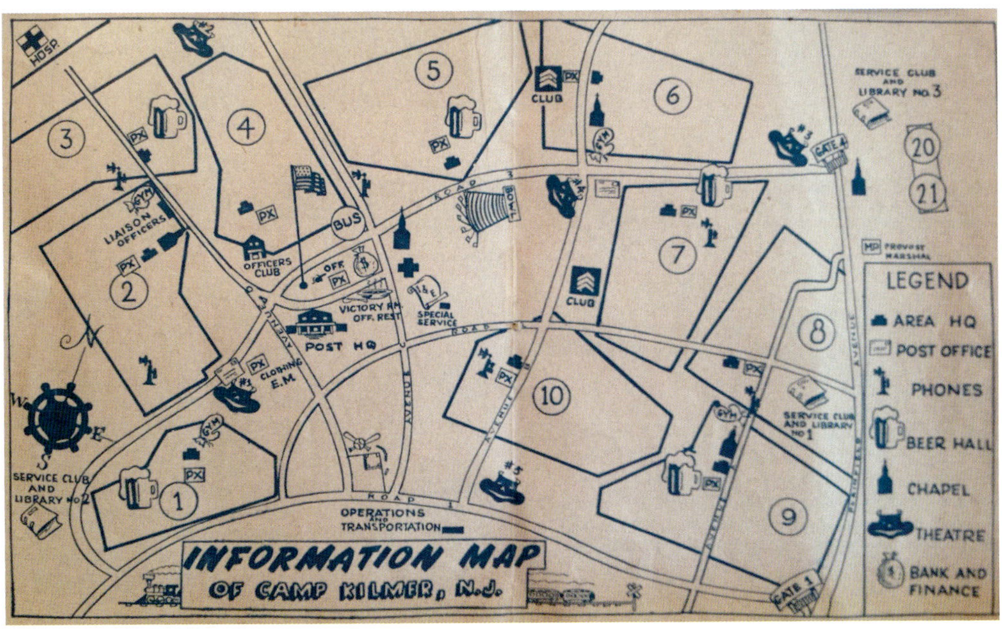

Sketch map showing the general layout of Camp Kilmer and its ten barrack areas as configured during late World War II. A small portion of the camp was set aside beginning in May 1945 for use as a POW camp and ISU.

Source : Camp Kilmer booklet published by Special Service Branch, U.S. Army, Camp Kilmer, NJ, 1945

Port Terminals, New York ISU

NYU-32-1-1

NYU-32-1-5

NYU-32-1-10

NYU-32-1-25

No.	Denomination	Paper	Ink	Serial	Block	Circ.	Unc
NYU-32-1-1	1 cent	pale orange	blue	red	A	25	40
NYU-32-1-5	5 cents	pale green	blue	red	A	35	55
NYU-32-1-10	10 cents	buff	blue	red	A	55	85
NYU-32-1-25	25 cents	gray	blue	red	E	90	150

Raritan Arsenal ISU

NJU-30-1-1

NJU-30-1-5

NJU-30-1-10

NJU-30-1-25

NJU-30-2-1

NJU-30-2-5

NJU-30-2-10

NJU-30-2-25

NJU-30-2-50

No.	Denomination	Paper	Ink	Serial	Block	Circ.	Unc.
	Exchange (blue)						
NJU-30-1-1	1 cent	white	blue	red	B	25	40
NJU-30-1-5	5 cents	rose	blue	red	B	35	55
NJU-30-1-10	10 cents	buff	blue	red	B	55	85
NJU-30-1-25	25 cents	pale gray	blue	red	B	90	150
	Exchange (black)						
NJU-30-2-1	1 cent	white	black	red	F	25	40
NJU-30-2-5	5 cents	pink	black	red	F	35	55
NJU-30-2-10	10 cents	buff	black	red	F	55	85
NJU-30-2-25	25 cents	pale gray	black	red	F	90	150
NJU-30-2-50	50 cents	pale gray	black	red	F	125	200

Unknown Chits

A few chits that have been found that don't have the name of any particular camp. One set of denominations of 1 cent through 50 cents has "P.O.W." and "Canteen" stamped on a chit that has been printed with the words Coupon Check, the denomination, No., and the usual phrase "not good if detached." There is a serial number printed in red. These chits have been found as individual chits and in sheets of five. They are perforated and have different serial numbers. Some collectors have thought of these chits as fantasies, but there is also a theory that they were used as generic temporary chits until chits with the camp location could be printed. For now they remain a mystery and, hopefully, new information about them will be discovered.

Shown below is another POW chit with no base to identify its location. This chit is a different size and style than the typical chit. It was found with another batch of known chits, so it is likely to be genuine rather than a fantasy. Again, maybe a collector reading this book will provide an answer.

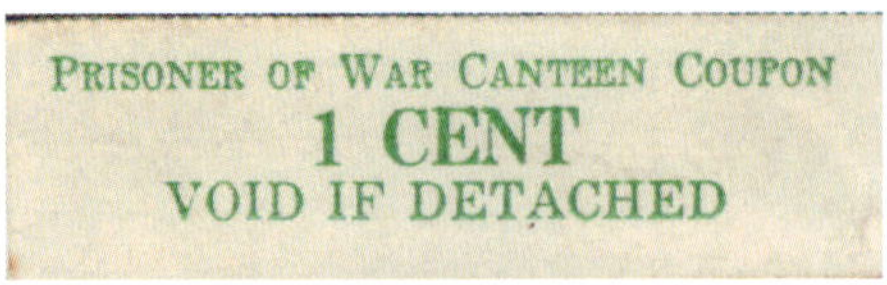

LIST OF CHIT SERIAL NUMBERS BY THE AUTHORS

The following list of serial numbers was compiled by the authors in hopes that other collectors would add to the list to create a database of all known chit serial numbers.

State	Camp	Serial #	Denomination	Notes
AL	Aliceville			
		2503G	2,5, and 10 cents	
		40376A	2 and 10 cents	
	Camp Rucker	none	2 cents	Laser Fantasy black ink
		16460A	2 cents	Laser Fantasy red ink
		6804	5 cents	Laser Fantasy red ink
	Ft. McClellan	(2) 46236C	5 cents	
		21642C	$5 booklet	
	Opelika	85108	10 cents	
		85363	1 cent	
		88134	1 cent	
		5002B	1 cent	
	Opelika Int. Camp	5002B	1, 5, and 10 cents	
AK	Excursion Inlet	A1783	1 cent	
AZ	Davis-Monthan	626A	1 and 5 cents	
	Florence	136351	1 cent	
		138048	1 cent	
		118773	5 cent	
		125411	5 cents	
		A03871	1 cent	U.S. Military Personal POW Camp Exhange
		A67294		booklet no value on cover

State	Camp	Serial #	Denomination	Notes
AZ	Florence	A130343	1 cent	
		A130376	10 cents	
		B170111	10 cents	
		B158224	5 cents	
		B169674	1 cent	
		B179277	1 cent	
		138086	5 cent	
		C194742	1cent	
	Papago Park	626A	1 and 10 cents	$1 booklet
		10001D	2 and 10 cents	
		16001E	25 cents	
		1251G	2, 5, and 10 cents	
		110001H	1, 5, and 10 cents	$1 booklet
		31251J	1, 5, and 10 cents	
AR	Camp Chaffee	87788	2 cents	
	Dermott	A03511	1 cent	
		A00334	25 cents	
		A01990		$5 booklet
		B01596	50 cents	
	Jerome	B004033	1 cent	
		B008281	1, 5, 10, 25, and 50 cents	
		D004892	1 cent	
		D017683	1 cent	
		D020003	1, 5, 10, 25, 50, and $1	$20 booklet
		B004892	1 cent	
		C07866		$15 booklet
		C07957		$15 booklet
		D015256	50 cents	

State	Camp	Serial #	Denomination	Notes
AR	Camp J. T. Robinson	7599B	1 cent	
		32828E	1 cent	generic merchandise chit from POW
		(B) 29793	1 cent	exchange
		(B) 29987	1 cent	exchange
		22648L	10 cents	
	Monticello	58931	1 cent	
		17929	5 cents	
CA	Camp Anza	A00295	10 cents	
	Camp Beale	74605	1 cent	
		A5002	5, 10, 25, and 50 cents	
	Camp Cooke	2922	1 and 5 cents	
		3708	1 and 5 cents	
		8092	10 cents	
	Camp Haan	B32498	5 cents	
		B50800	1 cent	
		B51137	1 cent	
		B51277	1 and 10 cents	
		B51318	1 cent	
	Camp Roberts	13486	5 cents	
		45365	1 cent	
		17702	10 cents	
	Camp Stoneman	A23	1, 5, and 10 cents	
		A1149	1 cent	
		C587	1, 5, and 10 cents	
		C3581	5 cents	
		C4610	5 cents	
		C7208	1 cent	
		C698	1 cent	
		C2112	5 cents	

State	Camp	Serial #	Denomination	Notes
CA		C5091	10 cents	
		C587	25 cents	
	DeWitt Gen. Hosp.	2503E	1, 5, 10, and 25 cents	
	Ft. Ord	(W) 10884	1 cent	
		(W) 25087	1, 5, 10, and 25 cents	$3 booklet
		(WC) 2885	5 and 10 cents	$2 booklet
		10092C	1 cent	Pamaboco coupon check
	Stockton	20015	1, 5, 10, 25, and 50 cents	
		(9A) 39554	25 cents	
		B10009	1, 5, 10, and 25 cents	
		B20014	5, 10, and 25 cents	
		B30006	1 cent	
		C43506	5 and 10 cents	
		(K) 17384	1 cent	
	Pomona & SCU 1894	C03420	1 cent	
		A04007	1 cent	
		A04742		$2 booklet
CO	Camp Carson Int. Camp	6512	5 and 10 cents	
	Camp Carson	1185F	1 cent	
		87K	1 cent	
		1744K	1 cent	
		2723L	1 cent	
	Greeley	20010	1, 5, 10, and 25 cents	
		(J) 355A	2 cents	
		(J) 1951D	5 cents	
		(J) 4013A	5 cents	
		(D) 12889A	1 cent	
		33831B	2 cents	
		29403C	2 cents	

State	Camp	Serial #	Denomination	Notes
CO	Greeley		29567C	10 cents
		30964C	5 cents	
		30964D	5 cents	
	Trinidad Int. Camp	33087	1 cent	
	Trinidad	A02411	1 cent	
DE	Ft. Dupont	1477C	1 cent	
		14219C	1 cent	
		2802F	5 cents	
		2901G	1 cent	
		259H	5 cents	
		2968G	5 cents	
	Ft. Dupont 1265th SCU	7368B	1 cent	
FL	Camp Blanding	25015	5 cents	
		18186C	1 cent	
		18804C	1 cent	
		23159C	1 and 5 cents	$5 booklet
		53596C	5 cents	
		61266C	10 cents	
		7004D	1 cent	
		7508D	1 cent	
		16291D	1 cent	
		47563F	5 cents	
		52667F	5 cents	
		42820F	5 cents	
		61276F	1 cent	
		61307C	10 cents	
		61236C	25 cents	

State	Camp	Serial #	Denomination	Notes
FL	Camp Gordon Johnston	27A	1 cent $5 booklet	
GA	Ft. Benning	52552B	1 cent	
		(3) 591K	1, 5, 10, and 25 cents	$5 booklet
		(8) 18384J	10 cents	
		(8) 5772K	25 cents	
		(10) 10465D	25 cents	
		(10) 10604D	1 cent	
		(11) 16207D	10 cents	
		(12) 24151D	25 cents	
		47738C		$5 booklet
	Camp Wheeler	(3) 34811A	1 cent	
		(5) 6567B	1 cent	
		(6) 1430B	5 cents	
		(7) 3321B	1 cent	
		20820C	10 cents	
		12284D	1 cent	
		14635D	1 cent	
		2123E	5 cents	
		12510E	1, 5, 10, and 25 cents	$3 booklet
		7629F	1 cent	
		20303G	5 cents	
HI	Central Pacific Base	5381B	1, 5, and 10 cents	$2 booklet
ID	Rupert	none	1 cent	
		1197	1 cent	
		13709	1, 5, 10, and 25 cents	
		16431	1 cent	

State	Camp	Serial #	Denomination	Notes
ID	Rupert	19449	1 cent	
		20935	1 cent	
		26719	1 cent	
		34844	1, 5, 10, and 25 cents	
		35000	5, 10, and 25 cents	
		6618	5 cents	
	Farragut	11544	1 cent	
IL	Camp Ellis	30869	1 cent	80 cents booklet
		48405	1, 5, and 10 cents	80 cents booklet
		42912	1 cent	
		(3) 35209B	1 cent	$3 booklet
		5491B		$3 booklet
		34609C	2 cents	
		17666B	5 and 10 cents	$3 booklet
		12867D	25 and 50 cents	
		6050C	2 cents	
		16211A		$1 booklet
		14343C		$4 booklet
		8006D		$10 booklet
	Camp Grant	8347A	5 cents	$1 booklet
		8948A	1 cent	
		18103A	1 cent	
		5491B		$3 booklet
		6050C	2 cents	
		6349C	5 cents	$3 booklet
		37396C	1 cent	
		22231E	1 cent	
		32912E	1 cent	
		35132E	25 cents	
		34617E	1, 5, 10, and 25 cents	$5 booklet

State	Camp	Serial #	Denomination	Notes
IL	Camp Grant	35132E	5 and 25 cents	
		27252E	1 cent	
		27274E	1 cent	
		28021E	1 cent	
		29073E	1 cent	
		33021E	1 cent	
		41713E	2 cents	
		3152F	1 cent	20 cent booklet
		2211G	1 cent	40 cent booklet
		33192E	2 cents	
	Ft. Sheridan	21811A	5 cents	
		21956A	1 cent	
		22049A	1 and 5 cents	
		25479A	1 cent	
		29481A	1 cent	
		29552A	1 cent	
		29610A	5 cents	
		33515A	5 cents	
		6888H	1 cent	
		4665B	1 cent	
		7243B	1 cent	
		9109B	10 cents	
		9127B	5 cents	
		11050B	5 cents	
		7072C	1 cent	
		7557C	10 cents	
		392F	1 cent	
		50685H	1 cent	
		2209J	2 dollars	
		2232J	1 dollar	
		2300J	25 cents	
		2302J	1 dollar	

State	Camp	Serial #	Denomination	Notes
IL	Ft. Sheridan	2323J	50 cents	
IN	Atterbury Int. Camp	5114A	1 and 5 cents	
		5006C	1, 5, and10 cents	
	Atterbury 1537th SCU	52807A	5 and 10 cents	
		52934A	1 cent	
		(II) 2908D	1, 5, 10, and 25 cents	$5 booklet
		(II) 26405D	25 cents	
		(II) 39243D	1 cent	
		(II) 26405D	25 cents	
	Atterbury 1560th SCU	(V) 45156C	1 cent	
		(V) 52745C	1 and 5 cents	
		52834C	1 and 5 cents	$5 booklet
	Ft. Benjamin Harrison	29203	1, 5, and 10 cents	
IA	Algona	30501A	1, 5, and 10 cents	
		90004A	1, 5, and 10 cents	$1 booklet
		61318A	5 cents	
		102501A	1, 5, and 10 cents	$1 booklet
		107822A	10 cents	
		51391C	1, 5, 10 cents	
		56752C	10 cents	
		7501D	1, 5, 10, 25, and 50 cents	
		25077D	25 cents	
		25126D	1 cent	
		29979D	25 cents	
	Clarinda	0992	1 cent	
		6982B	1 cent	
		2839D	1 cent	
KS	Camp Phillips	38880		receipt for $3 booklet
		38180	1 cent	

State	Camp	Serial #	Denomination	Notes
KS	Ft. Riley	30876	1 and 5 cents	$1 booklet
		37001	1 and 5 cents	
		51251	1 and 5 cents	
		56600	$1 booklet	
		2001A	1 and 5 cents	
		50001A	1 and 5 cents	
		0031	10 cents	
KY	Camp Campbell	5529	1, 5, 1, and 25 cents	$5 booklet
		37641A	5 cents	
	Ft. Knox	10769	1 cent	
		15282C	1 cent	
		30001C	1, 5, 10, and 25 cents	
		30002C	1 and 5 cents	
		52261B	5 cents	
		10772	5 cents	
		5198B	1 and 5 cents	
		59116		camp theatre ticket
LA	Camp Claibourne	12393B	5 cents	
		8317C	10 cents	
		100436C	10 cents	
	Camp Polk	C04522		$5 booklet
		27869	5 cents	
		C07590	1 cent	
		6137D	5 cents	Exchange
		2972P	1 cent	Exchange
	Camp Livingston	A22098	1 cent	
		B01755	1 cent	
		B38836	1 and 10 cents	

State	Camp	Serial #	Denomination	Notes
LA	Camp Livingston	A297851	1 cent	
	Camp Livingston			
	Int. Camp	A02259	1 cent	
	Camp Plauche	1932	5 cents	trade coupon
	Ruston	7001A	1, 5, and 10 cents	
		1501B	1 cent	
		39751F	1, 5, 10, 25, and 50 cents	
		132G	25 cents	
		2225H	10 cents	
		2240H	1 cent	
		3751H	1, 5, 10, 25, and 50 cents	
		2027S	5 cents	
ME	Houlton	5446F	1 cent	
		(2) 58766G	2, 5, and 10 cents	
		5446F	1 cent	
		7178E	1 cent	
		1220E	5 cent	
MD	Ft. George G. Meade	401A	1 cent	
		9936A	5 and 10 cents $	2 booklet
		35889A	10 cents	
		898813A	1 cent	
		7972B	10 and 25 cents	
		19381B	1, 2, 5, 10 cents	$10 booklet
		20150B	1, 5, 10, and 25 cents	$10 booklet
		7251C	1, 5, and 10 cents	
		38938C	5 cents	
		14325D	1, 5, 10, and 25 cents	$3 booklet
		15495E	5 cents	
		44949E	2 cents	

State	Camp	Serial #	Denomination	Notes
MD	Ft. George G. Meade	127711E	2, 5, 10, and 25 cents	$5 booklet
		147518E	1 and 2 cents	
		60851E	5 cents	
		7725B	10 cents	
		117439E	25 cents	
MA	Ft. Edwards	577C	2 cents	
		Ft. Devens	81934 1 cent	
		0728	1 cent	
		2777	1, 5, and 10 cents	
		1023	10 cents	
		24097	5 cents	post exchange
		46871	1 cent	
		7281	1 cent	
		35690	1 cent	
MI	Ft. Custer	2589D	1 cent	$5 booklet
		48638E	1 cent	
		161G	1 and 2 cents	$5 booklet
		919G	1 and 2 cents	$5 booklet
		925G	1 cent	
		1285G	1 and 2 cents	$5 booklet
		1387G	2 and 5 cents	$5 booklet
		12813G	1 cent	
		18619N	1 cent	
		22074N	5 cents	
		33729M	5 cents	
	Owossa	none	1 cent	

State	Camp	Serial #	Denomination	Notes
MS	Camp McCain	19820	5 cents	
		B 34502	5 cents	
		C 17367C	1, 5, 10, 25, and $1	
		C 17568C	25 cents	
		27117J	1 and 5 cents	
	Camp Shelby	9020D	1, 2, 5, 10, and 25 cents	
	Como	25376A	1, 5, 10 cents	$1 booklet
		50251E	1, 5, 10, and 25 cents	
	Como Int. Camp	5626A	1 cent	
		11251E	25 cents	
MO	Camp Clark Int. Camp	1001A	1, 5,and 10 cents	
		1001C	1 cent	
	Camp Clark Int. Camp	2501	1 and 10 cents	
		39938	1 cent	
		28001A	1, 5, 10, and 25 cents	$3 booklet
		8270S	5 cents	
	Camp Crowder	73797B	1 cent	
		149888B	1 cent	
	Ft. Leonard Wood	2931	1 cent	
		27073	25 cents	
	Weingarten Int. Camp	2276	1 cent	
		2290	1 cent	
		2122	5 cents	
		2318	10 cents	
	Weingarten Int. Camp	10173C	1 cent	
		59269C	25 cents	
NE	Atlanta	6883	1 cent	
		C 4095	1 cent	
		C 8265	1 cent	

State	Camp	Serial #	Denomination	Notes
NE	Atlanta	C 12428	1, 5, 10, 25, and 50 cents	$10 booklet
	Ft. Crook	44167BL	1 cent	
	Ft. Robinson Int. Camp	4004A	1, 5, and 10 cents	
	Ft. Robinson	21251A	1, 5, 10 cents	$1 booklet
		2501D	25 cents	
		3501E	1, 5, 10, and 25 cents	
	Indianola	45128	1 cent	
		45134	1 cent	
		45159	1 cent	
	Scottsbluff	19629B	2 and 5 cents	
NJ	Kilmer	126A	1, 5, and 10 cents	
		1157F	1, 5, 10, 25, and 50 cents	
		1244F	1, 5, 10, and 25 cents	$10 booklet
	Kilmer Type 2	(1) 376B	1, 5, 10, and 25 cents	
		(1) 3201B	1, 5, 10, and 25 cents	
		(1) 6251B	1 and 5 cents	$2 booklet
		(1) 376C	1, 5, 10, and 25 cents	$3 booklet
	Kilmer Camp Canteen	(1) 626C	1, 5, 10, and 25 cents	
		(4) 876C	1, 5, and 10 cents	$3 booklet
	ASF Belle Meade	9801E	1, 5, 10, and 25 cents	
		11501F	1, 5, 10, 25, and 50 cents	
		11812F	1, 5, 10, 25, and 50 cents	
	Belle Meade ISB	1504F	1, 5, 10, 25, and 50 cents	
		4501F	1, 5, 10, 25, and 50 cents	$10 booklet
	Raritan Arsenal ISB	13752B	1, 5, 10, and 25 cents	
		10126B	1 and 25 cents	
		3004F	1, 5, 10, 25, and 50 cents	
	Ft. Dix	112448A	1 cent	
		123724A	25 cents	

State	Camp	Serial #	Denomination	Notes
NJ	Ft. Dix	(3) 72956	5 cents	
		32533B	1, 5, 10, and 25 cents	
NM	Lordsburg	10255	1 cent	
		2159	1 cent	
		1251B	1 and 5 cents	
		2502E	1, 5, and 25 cents	
		10001F	1, 5, 10, 25, and 50 cents	$10 booklet
	Roswell	1251C	1, 5, 10, and 25 cents	
		16967C	1 cent	
		18408C	25 cents	
		1251D	25 cents	
		50387D	1 cent	
		50795D	1, 5, and 10 cents	
		253E	1 cent	
		34220P	5 cents	
		46049P	25 cents	
	Roswell Int. Camp	8251C	1, 5, 10, and 25 cents	
NY	Ft. Niagara	5175B	1 cent	
		9439C	1 cent	
		3001F	1, 5, and 10 cents	
		6751F	1, 5, and 10 cents	
		17001F	1, 5, and 10 cents	
		24251F	1, 5, and 10 cents	$2 booklet
		26251F	1, 5, and 10 cents	
		28251F	1, 5, and 10 cents	$2 booklet
	Pine Camp 1277th S.C.S.U.	3001A	1 and 5 cents	
	Port Terminals	126A	1 cent	
		4626A	1 and 5 cents	$1 booklet

State	Camp	Serial #	Denomination	Notes
NY	Port Terminals	5376E	1, 5, 10, and 25 cents	$5 booklet
	Halloran Gen. Hospital	4243	1 cent	
		4418	5 and 10 cents	
	Pine Camp 1209th SCU	18H	5 cents	
	Sampson 1215th SCU	190D	1 cent	
NC	Camp Mackall	1398B	1 cent	
		3545C	1 cent	
		1422B	5 cents	
		1415B	5 cents	
		1520B	2 cents	
		1528B	2 cents	
		3285C	10 cents	
		4912C	10 cents	
		1451D	5 cents	
		1594D	25 cents	
		2658D	25 cents	
	Ft. Bragg	19962B	1 cent	
OH	Camp Perry	231281	1 cent	
		269756	5 cents	
		270754	10 cents	
		266552	10 cents	
		269752	25 cents	
		11041A	1 cent	
OK	Alva Int. Camp	626B	5 and 10 cents	
		10780B	5 cents	
		13970B	1 cent	

State	Camp	Serial #	Denomination	Notes
OK	Alva	251C	1, 5, 10, and 25 cents	
		52126C	1, 5, 10 cents	$3 booklet
		23027E	1, 5, 10, and 25 cents	$5 booklet
		29455E	1 cent	
		66188E	1, 5, 10, and 25 cents	$5 booklet
	Glennan Gen. Hospital	1251	1 and 10 cents $1 booklet	
		126C	1, 5, 10 cents $3 booklet	
		502C	1, 5, 10, and 50 cents	
		1751C	1, 5, 10 cents	
		5501C	1, 5, 10, 25, and 50 cents	
		1002J	1, 5, 10, and 50 cents	
		1004J	50 cents	
	El Reno	2404D	1 cent	
		0709	1 cent	
	McAlester	2943	1 cent	
		11675	1 cent	
		25771	10 cents	
		25862	1 cent	
		12490	1 cent	trade coupon
		103292	1, 5, 10, and 25 cents	trade coupon
	Oklahoma City Air Depot		8827B 1 cent	exchange
	Camp Gruber	42153	1 cent	
		C029362	1 cent	
		39896	1 cent	
		25768	2 cents	
	Ft. Sill	3134A	5 cents	
		26228C	5 cents	
	Pryor	A20003	1, 5, 10, and 25 cents	$5 booklet
	Tonkawa	42132	5 cents	trade coupon
		60200	1 cent	trade coupon

State	Camp	Serial #	Denomination	Notes
OR	Camp White	0448	1, 5, and 10 cents	
		59044		$5 empty booklet
		30007	1, 5, 10, and 25 cents	$5 booklet
		81264	1 and 5 cents	
	Camp Adair	A1897	1 cent	
PA	Camp Reynolds	35978B	2 cents	
		13698D	1, 2, 5, 10, and 25 cents	$10 booklet
	Olmsted Field	126E	1, 5, 10, and 25 cents	
		5004F	1, 5, 10, 25, and 50 cents	
	Valley Forge Gen. Hospital		1751B1 cent	spelling error Phocnixville
		1751B	1, 5, and 10 cents	
		256E	1, 5, 10, and 25 cents	$5 booklet
		2752E	1, 5, and 25 cents	
		2501F	1, 5, 10, 25, and 50 cents	$10 booklet
	Tobyhanna	2391	1 cent	
RI	Harbor Defense	(6) 00256	1, 5, and 10 cents	
		(6) 00646	1, 5, and 10 cents	$1 booklet
		(6) 00649	1, 5, and 10 cents	$1 booklet
SC	Ft. Jackson	(2) 30001	1 cent	
		(3) 42742	1 cent	
		155045C	5, 10, and 25 cents	
		158247C	1, 5, 10, and 25 cents	

State	Camp	Serial #	Denomination	Notes
TN	Crossville	504A	1, 3, and 5 cent	
		939A	3 cents	
		456B	1, 5, and 10 cents	$2 booklet
		501E	1, 5, 10, and 25 cents	
		502E	1, 5, 10, and 25 cents	$5 booklet
		1104F	1, 3, 5, 10, 25, and 50 cents	
	Crossville Int. Camp	3751Bb	1, 5, and 10 cents	
		5001B	1, 5, and 10 cents	$2 booklet
	Camp Forrest	5460B	1 cent Exchange	
		17751C	1, 5, 10, and 25 cents	$5 booklet
		37097C	1 cent	
		37173D	10 cents	
		35857D	1 cent	
		38731D	1 cent	
		41251D	1, 5, 10, and 25 cents	$5 booklet
		63328D	5 cents	
		71631D	10 cents	
		131251D	1, 5, 10, and 25 cents	$5 booklet
		147207D	1 cent	
		147208D	25 cents	
		150027D	1 cent	
		198813D	1 cent	
		209826D	1 cent	
		1789E	1 cent	
		16868E	1 cent	
		101251E	1, 5, 10, and 25 cents	
		668F	1, 5, and 10 cents	
		1001F	50 cents	
		16001F	1, 5, 10, 25, and 50 cents	$10 booklet
		37001F	1, 10, 25, and 50 cents	
		15562F	1 cent	
		44582F	1 cent	

State	Camp	Serial #	Denomination	Notes
TN	ASF Depot Memphis	251A	1 and 10 cents	
		4001A	1, 5, and 10 cents	$1 booklet
		6501A	1 and 5 cents	$1 booklet
		10501A	1 and 10 cents	$1 booklet
		18501A	1 cent	
		28751A	1, 5, and 10 cents	
		64601E	1, 5, 10, and 25 cents	
		58901E	1, 5, 10, and 25 cents	
TX	Camp Barkeley	126B	5 cents	
		1991B	10 cents	
		2501E	10 cents	
		5501E	10 cents	$5 booklet
	Camp Bowie	1528	5 cents	
		17395	10 cents	
		22841	5 cents	
		62501	1, 5, and 10 cents	
		37278	1, 5, and 10 cents	$3 booklet
		751B	1 and 5 cents	$1 booklet
	Camp Brady	13151	1 cent	expediant no official stamp
		16005	1 and 2 cents	expediant stamping
		2129A	1 cent	
		120B	2 cents	
	Camp Fannin	21773A	5 cents	
		22094A	25 cents	
		22112A	1 cent	
		22021A	10 cents	
	Camp Hood	5904F	1 cent	exchange in POW estate

State	Camp	Serial #	Denomination	Notes
TX	Camp Hulen	2626E	1 and 5 cents	
	Camp Howze	(6) 13001	1 cent	
	Camp Maxey	A02004	1, 5, and 10 cents	$1 booklet trade coupon
		2251A	1 cent	
		5391E	1, 5, 10, and 25 cents	
	Camp Swift	69314	1 cent	
		4376C	1, 5, 10, and 25 cents	
		19856C	5 cents	
		20244C	25 cents	
		75484E	1 cent	
		75361E	5 cents	
		64741E	10 cents	
	Camp Wolters	6582	1 cent	
		7175	25 cents	
		7301	10 cents	
		7322	50 cents	
		12921	1 cent	
		16070	1, 5, and 10 cents	
	Ft. Bliss	3507	1 cent	
	Ft. Clark	2165A	1 cent canteen	
		1618C	1 cent canteen	
	Ft. Crockett	8412	10, 25, and 50 cents	trade coupon
		8676	5 cents	trade coupon
	Ft. Sam Houston	9421B	1 cent	
	Hearne	5780	1 cent	
		17197	1 cent	
		35535		.40 receipt
		3501N	1 and 5 cents	.50 booklet
		11818N	1 cent	
		16313N	1 cent	
	Hereford	3101B	1 cent	

State	Camp	Serial #	Denomination	Notes
TX	Hereford	38509C	25 cents	
		855H	1 cent	
		1256J	1 cent	
	Huntsville	37626A	1, 5, and 10 cents	$1 booklet
		626B	1, 5, 10, and 25 cents	$3 booklet
		661C	10 cents	
		17751C	1, 5, 10, and 25 cents	
		14793E	1 and 5 cents	
	McLean	3723	1 cent	
		10546	1 and 10 cents	
		12436	1 cent	
		14029	1 cent	
		8098A	1 cent	
		11029	1 cent	
	Mexia	251A	1, 5, and 10 cents	
		626A	1, 5, and 10 cents	
		18626A	1, 5, and 10 cents	
		35692C	1 cent	
		14626D	1, 5, 10, and 25 cents	
		61001E	1, 5, 10, 25, and 50 cents	$10 booklet
	Mexia Int. Camp	2251A	1, 5, and 10 cents	$1 booklet
		3251C	1, 5, 10, and 25 cents	
	North Camp Hood	2863A	5 cents	
		3008A	1 cent	
	Randolph Field	31163E	1 cent	Exchange in POW estate
UT	Ogden	197052	1 and 5 cents	
		197041	10 cents	
		75310	1 cent	

State	Camp	Serial #	Denomination	Notes
VA	Camp Ashby	252A	1, 5, and 10 cents	
	Camp Ashby Branch			
	Camp	810D	1, 5, 10, 25, and 50 cents	
		3666H	50 cents	
	Camp Patrick Henry	251A	1, 5, and 10 cents	
		3001A	1 cent	
		3504C	1, 5, and 25 cents	
		6626C	5, 10, and 25 cents	$3 booklet
		126G	1, 5, and 10 cents	
	Camp Peary	(2) 1558F	5 cents	
	Camp Pickett	29878	1 and 5 cents	
		12654C	1 cent	
	Ft. Eustis	9704C	1, 5, 10, and 25 cents	
	Front Royal	22799	1 cent	
WA	Ft. Lewis	6859	1, 2, 5, and 10 cents	
		31761	5 cents	
		31946	25 cents	
		31975	1 and 2 cents	
		32565	10 cents	
		50871	1 cent	
		77944	1 cent	
		92927	2 cents	
		13094A	10 cents	
		13127A	25 cents	
		13177A	1, 2, and 5 cents	
		11442	1 cent	
	Ft. Lawton	0041	5, 10, and 25 cents	

State	Camp	Serial #	Denomination	Notes
WV	White Sulphur Springs	108260B	10 cents	
WI	Billy Mitchell Field	132082A	1, 5, and 10 cents	
	Camp McCoy	74625A	1, 5, and 10 cents	$1 booklet
		1086B	1 cent	Post Exchange
		3492B	5 cents	Post Exchange
WY	Douglas	4274	10 cents	
		626A	1 and 10 cents	$1 booklet
		5001A	1, 5, and 10 cents	$1 booklet
		11251A	10 cents	
		36126A	1 cent	
		13326B	1 cent	
		411A	1 cent	

BIBLIOGRAPHY

Campbell, L. K. *Prisoners of War and Concentration Camps of the 20th Century*. Port Clinton, OH: BNR Press, 1993.

Donn, A. E. *World War II Prisoner of War Scrip of the United States*. Iola, WI: Krause Publications, 1970.

Lt. Col. Lewis, G. G, & Capt. John Mewha. *History of Prisoner of War Utilization By The United States Army 1776-1945*. Washington, DC : Department of the Army, 1955

Pick, A. *Die Kantinenschecks der Deutschen Kriegsgefangenen in den U.S.A.* Verlag Cologne, 1953

Schwan, C. F. & J. E. Boling. *World War II Remembered, History in Your Hands, a Numismatic Study*. Port Clinton, OH: BNR Press, 1995.

Slabaugh, A. *Prisoner of War Monies and Medals*. Hewitt's Numismatic Information Services. Chicago 1965.

Index

WAR DEPARTMENT
OFFICE OF THE PROVOST MARSHAL GENERAL

PRISONER OF WAR CAMP LABOR REPORT

Reports Control Symbol MGA 48 (R1)

PMG OFFICE USE ONLY

R		1	
E		2	
C		3	

CAMP: DOS PALOS POW BRANCH CAMP, Box 130-B. RFD #1
Firebaugh, California

PERIOD ENDING: 12 February 1946

BASE CAMP: PRISONER OF WAR CAMP FORT ORD, CALIFORNIA

NATIONALITY: German

Part I. TOTAL MAN-DAYS AND LABOR AVAILABILITY

1. TOTAL MAN-DAYS OF ALL PRISONERS IN THIS CAMP DURING THIS PERIOD:	OFFICERS	NCO'S	PRIVATES	TOTAL
	---	810	1152	1962

NONWORKERS		WORKERS	
2. REASON FOR NONWORKER STATUS:		3. WORK ASSIGNMENT STATUS:	
(a) PHYSICALLY UNABLE TO WORK		(a) WORKED	245
(b) TEMPORARY MEDICAL REASONS		(b) ASSIGNED, DID NOT WORK	
(c) DISCIPLINARY ACTION		(c) NOT ASSIGNED TO WORK	
(d) NONWORKING OFFICERS		TOTAL, WORKED OR AVAILABLE	245
(e) NONWORKING NCO'S		4. REST DAYS OF WORKERS	25
(f) OTHER REASONS (*explain*)	1692		
TOTAL MAN-DAYS, NONWORKERS	1692	TOTAL MAN-DAYS, WORKERS	270

Part II. MAN-DAYS WORKED, BY PROJECTS

PROJECT CODE	SHORT TITLE	TOTAL MAN-DAYS
	PW CAMP OVERHEAD	
10	COMPANY AND STOCKADE, UNPAID	59
11	COMPANY AND STOCKADE, PAID	109
12	PW CANTEEN	
13	PW OFFICERS' ORDERLIES, COOKS	
14	PW MEDICAL	11
15	CAMP HEADQUARTERS	66
16	CAMP PW FUND	
19	OTHER PW CAMP (*explain*)	
	SUBTOTAL	245
	SERVICE COMMAND ACTIVITIES	
21	SVC POST CONTRACT WORK	
22	SVC ACTIVITIES, THIS POST	
23	SVC ACTIVITIES, OTHER POSTS:	
	SUBTOTAL	
	TECHNICAL SERVICES ACTIVITIES	
	SUBTOTAL	
	ARMY AIR FORCES ACTIVITIES	
41		

PROJECT CODE	SHORT TITLE	TOTAL MAN-DAYS
	CONTRACT WORK	
51	AGRICULTURE	
52	PULPWOOD, LOGS, AND LUMBER	
53	MINING AND QUARRYING	
54	CONSTRUCTION	
55	FOOD PROCESSING	
56	OTHER MANUFACTURING	
57	TRANSPORTATION	
58	TRADE	
59	OTHER NONGOVERNMENTAL	
60	PUBLIC CONTRACT (*explain*)	
	SUBTOTAL	
	NAVY WORK	
71		
	RECAPITULATION	
	PW CAMP OVERHEAD	245
	SERVICE COMMAND ACTIVITIES	
	TECHNICAL SERVICES ACTIVITIES	
	ARMY AIR FORCES ACTIVITIES	
	CONTRACT WORK	
	NAVY WORK	
	TOTAL	245

* Must equal Part I, Item 3a.

WD AGO FORM 19-21, 1 JUL 1945. Replaces WD PMG Form 27, rev., and WD AGO 19-21 (1 Apr 1945), which are obsolete.

16—45110-1

Front and back (next page) of the final Prisoner of War Camp Labor Report from Dos Palos Branch Camp to the U.S. War Department, Provost Marshal General. Source : United States National Archives

Part III. REMARKS

Part I 8f

1692 waiting for Repatriation

CAMP CLOSED 12 February 1946

RECEIVED 15 ... 1946 P.M.G.O. M. & R.

DATE	SIGNATURE	RANK AND TITLE
12 Feb. 1946	Roy J. Stanfill	ROY J. STANFILL CAPT. INF.

INSTRUCTIONS

1. For full instructions see pars. 7 to 10, Chap. 7, and Appendix C, TM 19-500.

2. This report replaces WD AGO Form 19-21 (1 Apr 45) and WD PMG Form 27, rev.

3. Note that the report will require daily classification of prisoners as "workers" and "nonworkers." Prisoners on *essential* unpaid labor are workers and will be reported in Item 3*a*, Part I, and under Project Code 10, Part II.

4. Rest days are reported only for prisoners who are classified as workers. For each calendar week the number of rest days reported in Section 4 will equal approximately one-seventh of the total man-days (including rest days) of workers shown on the line following Section 4.

5. Before mailing this report, verify the addition, and the consistency of the following items: (*a*) Total man-days shown in Part I, Section 1 must equal the sum of man-days for nonworkers, Section 2, plus workers, Sections 3 and 4. (*b*) Total man-days worked, Item 3*a*, must equal total man-days shown in the recapitulation of Part II.

6. Where installation names are required, i. e., for service command activities at other posts, for technical services activities, for Army Air Forces activities, and for Navy work, identify the using installation by name. Do not give the name of an individual or subordinate agency thereof.

7. Prisoners on service command activities at technical services installations will be reported under project codes 10 to 19 for prisoner of war camp overhead, and project codes 22 or 23 for other service command activities.

8. Submit 2 reports each month: One for the period from 1st to 15th, inclusive, the other for period 16th to end of month. Mail one copy direct to Office of The Provost Marshal General, Washington 25, D. C., Attention: Prisoner of War Operations Division, to reach that office within 6 days from end of reporting period.

16—45116-1 U. S. GOVERNMENT PRINTING OFFICE

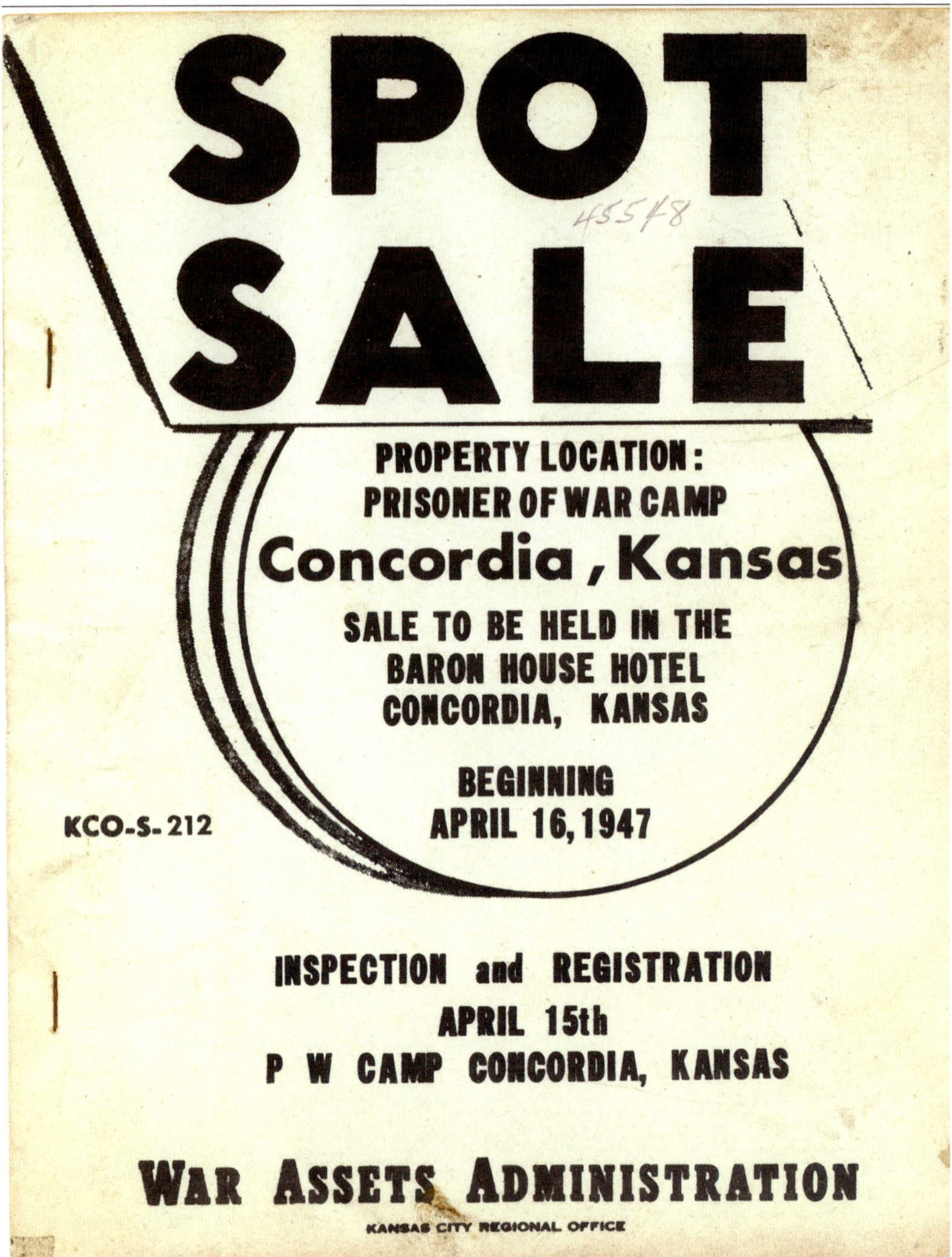
SPOT SALE

PROPERTY LOCATION:
PRISONER OF WAR CAMP
Concordia, Kansas
SALE TO BE HELD IN THE
BARON HOUSE HOTEL
CONCORDIA, KANSAS

BEGINNING
APRIL 16, 1947

KCO-S-212

INSPECTION and REGISTRATION
APRIL 15th
P W CAMP CONCORDIA, KANSAS

WAR ASSETS ADMINISTRATION
KANSAS CITY REGIONAL OFFICE

Cover of the sale catalog for the contents of the POW camp at Concordia, KS. It is an illustration of what happened to the camps and explains why there is very little remaining of any of them today.

PMG
PAPER MONEY GUARANTY
Oklahoma, Prisoner of War Camp
SB1051 ND (1944-46) 1 Cent
S/N 103292 - McAlester
Strip of 5
65
EPQ
Gem Uncirculated
TRADE COUPON
1 Cent
103292
McAlester Prisoner of war Camp Canteen

DAS HAUPTLAGER UND SEINE NEBENLAGER
BASE CAMP AND BRANCH CAMPS

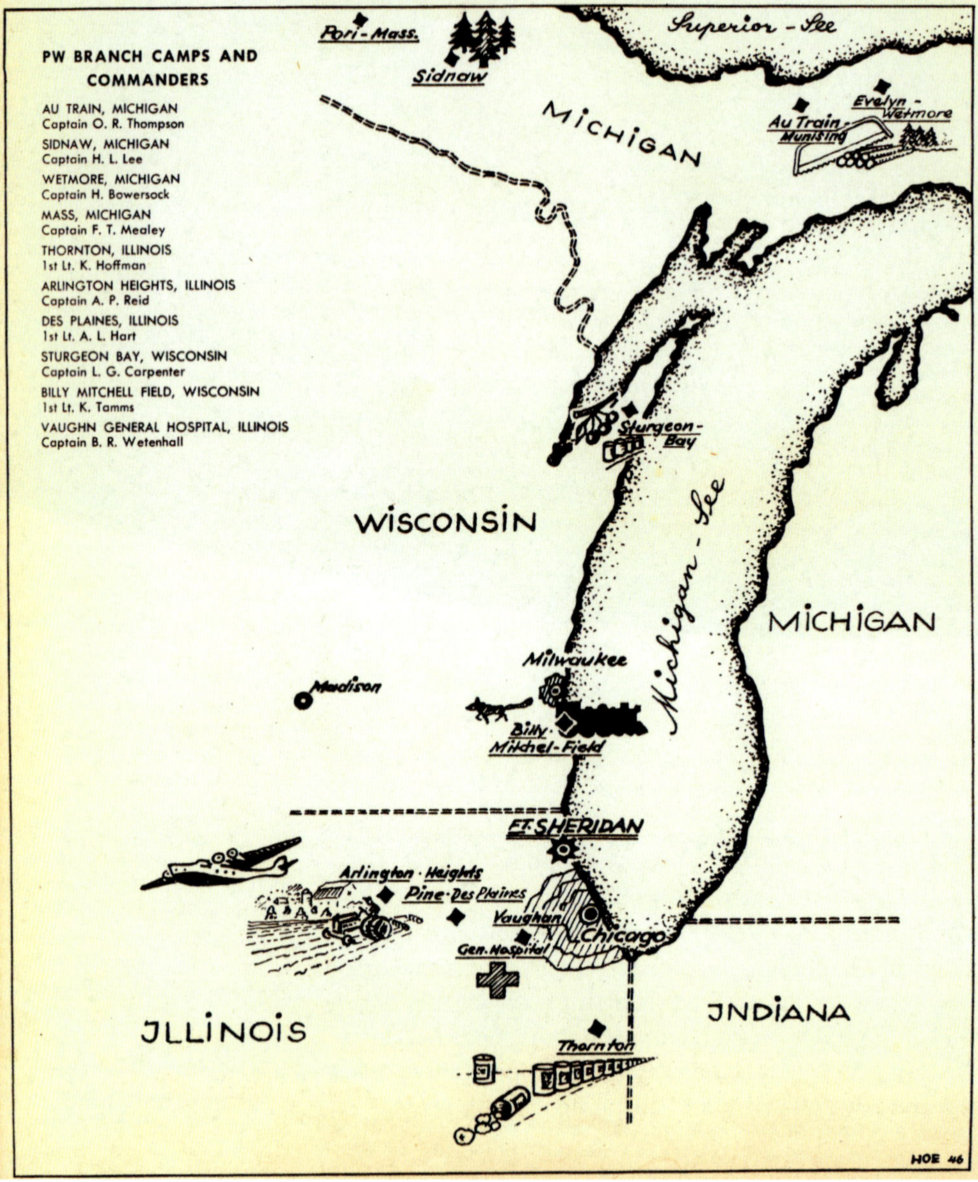

. . . and all this for 80 cents a day

. . . und das alles fuer 80 cents!